Smart Buildings Digitalization

T0312959

Smart Buildings Digitalization
Case Studies on Data Centers and Automation

Edited by
O.V. Gnana Swathika, K. Karthikeyan and
Sanjeevikumar Padmanaban

CRC Press
Taylor & Francis Group
Boca Raton London New York

CRC Press is an imprint of the
Taylor & Francis Group, an **informa** business

First edition published 2022
by CRC Press
6000 Broken Sound Parkway NW, Suite 300, Boca Raton, FL 33487-2742

and by CRC Press
4 Park Square, Milton Park, Abingdon, Oxon, OX14 4RN

First edition published by CRC Press 2022

CRC Press is an imprint of Taylor & Francis Group, LLC

ISBN: 978-1-032-14642-3 (hbk)
ISBN: 978-1-032-14643-0 (pbk)
ISBN: 978-1-003-24085-3 (ebk)

DOI: 10.1201/9781003240853

Typeset in Times
by codeMantra

Contents

Preface

Smart buildings are environmentally compliant and will enable people to realize a future low-carbon economy. This, in turn, will pave the way for integrating smart building applications with the information system. Also, smart building applications demand a multilayer software/hardware feature that contributes to attaining the best service with appreciable energy/cost reduction. Hence, it is crucial to understand and define the problem and provide intertwined software–hardware solutions. Robotics has numerous applications in smart buildings. Data centers form a pivotal role in the realization of smart buildings. IoT applications, such as data collection, public parking systems, smart metering, and sanitizer dispenser, are inexhaustible. Electric urban transport systems and effective electric distribution in smart cities are vital parameters of discussion. The extensive role of power electronics in smart building applications such as electric vehicles, rooftop terracing, and renewable energy integration is discussed. Case studies on automation in smart homes and commercial and official buildings are elaborated.

Five Salient Features:

1. Robotics in smart buildings
2. Role of power electronics in smart buildings
3. Data centers in smart buildings
4. IoT-based applications in smart buildings
5. Case studies: smart homes and commercial and office buildings

MATLAB® is a registered trademark of The MathWorks, Inc. For product information, please contact:

The MathWorks, Inc.
3 Apple Hill Drive
Natick, MA 01760-2098 USA
Tel: 508-647-7000
Fax: 508-647-7001
E-mail: info@mathworks.com
Web: www.mathworks.com

Editors

O.V. Gnana Swathika (Member '11–Senior Member '20, IEEE) earned a BE in electrical and electronics engineering at Madras University, Chennai, Tamil Nadu, India, in 2000; an MS in electrical engineering at Wayne State University, Detroit, MI, USA, in 2004; and a PhD in electrical engineering at VIT University, Chennai, Tamil Nadu, India, in 2017. She completed her postdoc at the University of Moratuwa, Sri Lanka. Her current research interests include microgrid protection, power system optimization, embedded systems, and photovoltaic systems.

K. Karthikeyan is an electrical and electronics engineering graduate with a master's in personnel management from the University of Madras. With two decades of rich experience in electrical design, he has immensely contributed toward the building services sector comprising airports, Information Technology Office Space (ITOS), tall statues, railway stations/depots, hospitals, educational institutional buildings, residential buildings, hotels, steel plants, and automobile plants in India and abroad (Sri Lanka, the UAE, and the UK). Currently, he is Chief Engineering Manager – Electrical Designs for Larsen & Toubro (L&T) Construction, an Indian multinational Engineering Procurement and Contracting (EPC) company. Also, he has worked at Voltas, ABB, and Apex Knowledge Technology Private Limited. His primary role involved the preparation and review of complete electrical system designs up to 110 kV. Detailed engineering stage covering by various electrical design calculations, design basis reports, suitable for construction drawings, and Mechanical Electrical and Public Health Services (MEP) design coordination. He is the point of contact for both the client and internal project team; leads and manages a team of design and divisional personnel, day-to-day interactions with clients, and peer reviews; manages project deadlines and project time estimation; and assists in staff appraisals, training, and recruiting.

Sanjeevikumar Padmanaban (Member '12–Senior Member '15, IEEE) earned a bachelor's degree in electrical engineering at the University of Madras, Chennai, India, in 2002; a master's (Hons) in electrical engineering at Pondicherry University, Puducherry, India, in 2006; and a PhD in electrical engineering at the University of Bologna, Bologna, Italy, in 2012. He was an Associate Professor at VIT University from 2012 to 2013. In 2013, he joined the National Institute of Technology, India, as a faculty member. In 2014, he was invited to be a visiting researcher at the Department of Electrical Engineering, Qatar University, Doha, Qatar, funded by the Qatar National Research Foundation (Government of Qatar). He continued his research activities with the Dublin Institute of Technology, Dublin, Ireland, in 2014. Further, he served as an Associate Professor in the Department of Electrical and Electronics Engineering, University of Johannesburg, Johannesburg, South Africa, from 2016 to 2018. From 2018 to 2021, he has been a faculty member in the Department of Energy Technology, Aalborg University, Esbjerg, Denmark. He is currently at Aarhus University, Herning, Denmark. He has authored over 300 scientific papers.

Dr. Padmanaban was the recipient of the Best Paper cum Most Excellence Research Paper Award from IET-SEISCON '13, IET-CEAT '16, IEEE-EECSI '19, and IEEE-CENCON '19 and five best paper awards from ETAEERE '16-sponsored Lecture Notes in Electrical Engineering – Springer. He is a fellow of the Institution of Engineers, India; the Institution of Electronics and Telecommunication Engineers, India; and the Institution of Engineering and Technology, UK. He is an editor/ associate editor/editorial board member for refereed journals, including the *IEEE Systems Journal, IEEE Transactions on Industry Applications, IEEE Access, IET Power Electronics, IET Electronics Letters,* and *Wiley-International Transactions on Electrical Energy Systems.* He is the subject editorial board member of *Energy Sources – Energies Journal, MDPI,* and the subject editor for the *IET Renewable Power Generation, IET Generation, Transmission and Distribution,* and *FACTS Journal* (Canada).

Contributors

Aashiq A
School of Electrical Engineering
Vellore Institute of Technology
Chennai, India

Aakash Aggarwal
School of Electronics Engineering
Vellore Institute of Technology
Chennai, India

Gangachalam Akula
School of Electronics Engineering
Vellore Institute of Technology
Chennai, India

Ananthakrishnan V
School of Electrical Engineering
Vellore Institute of Technology
Chennai, India

G.D. Anbarasi Jebaselvi
Department of ECE
Sathyabama Institute of Science
    and Technology
Chennai, India

S Angalaeswari
School of Electrical Engineering
Vellore Institute of Technology
Chennai, India

Umar Ahmad Ansari
School of Electrical Engineering
Vellore Institute of Technology
Chennai, India

Haniya Ashraf
School of Electrical Engineering
Vellore Institute of Technology
Chennai, India

R. Balakrishnan
Engineering Design Research Center
L&T Construction
Mumbai, India

V Berlin Hency
School of Electronics Engineering
Vellore Institute of Technology
Chennai, India

Dishant Bhagdev
L&T Construction
Mumbai, India

Parth Bhargav
School of Electrical Engineering
Vellore Institute of Technology
Chennai, India

A.M. Chithralegha
Engineering Design Research Center
L&T – Construction
Mumbai, India

S. Chowdhury
School of Electrical Engineering
National Institute of Technology Calicut
Kozhikode, India

J Christy Jackson
School of Computer Science and
    Engineering
Vellore Institute of Technology
Chennai, India

Milind Shrinivas Dangate
Chemistry Division, School of Advance
    Sciences
Vellore Institute of Technology
Chennai, India

Soham Deshpande
School of Electronics Engineering
Vellore Institute of Technology
Chennai, India

Kishore Eswaran
KPIT
Bangalore, India
and
University of Colorado Boulder
Boulder, Colorado

Xiao-Zhi Gao
School of Computing
University of Eastern Finland
Kuopio, Finland

O.V. Gnana Swathika
School of Electrical Engineering
Vellore Institute of Technology
Chennai, India

Gomathi V
Department of Electronics and
    Instrumentation Engineering
SASTRA University
Thanjavur, India

Naveen Kumar Gutha
Department of Electrical and
    Electronics Engineering
Andhra Loyola Institute of Engineering
    and Technology
Vijayawada, India

Jaanaa Rubavathay S
Department of Electrical and
    Electronics Engineering
Saveetha School of Engineering
Chennai, India

K Jamuna
School of Electrical Engineering
Vellore Institute of Technology
Chennai, India

V. Jayashree Nivedhitha
School of Electrical Engineering
Vellore Institute of Technology
Chennai, India

T. Kalavathi Devi
Department of EIE
Kongu Engineering College
Perundurai, India

Kanimozhi G
School of Electrical Engineering
Vellore Institute of Technology
Chennai, India

Priyanka Lal
School of Electronics Engineering
Vellore Institute of Technology
Chennai, India

Parth Mannan
School of Electrical Engineering
Vellore Institute of Technology
Chennai, India

V. Meenakshi
Department of Electrical and
    Electronics Engineering
Sathyabama Institute of Science and
    Technology
Chennai, India

Fahad Nishat
School of Electrical Engineering
Vellore Institute of Technology
Chennai, India

Siddharth Pandya
School of Electronics Engineering
Vellore Institute of Technology
Chennai, India

Abraham Sudharson Ponraj
School of Electronics Engineering
Vellore Institute of Technology
Chennai, India

B V A N S S Prabhakar Rao
School of Computer Science and
  Engineering
Vellore Institute of Technology
Chennai, India

Keerthi Priya Pullela
School of Electronics Engineering
Vellore Institute of Technology
Chennai, India

R. Rajapriya
Chemistry Division
School of Advanced Sciences
Vellore Institute of Technology
Chennai, India

Reena Monica P
School of Electronics Engineering
Vellore Institute of Technology
Chennai, India

K.D. Saha
School of Electrical Engineering
University of Southern California
Los Angeles, CA

C. Sai Aditya
School of Electronics Engineering
Vellore Institute of Technology
Chennai, India

P. Sakthivel
Department of EEE
Velalar College of Engineering and
  Technology
Thindal, India

C.M. Sarkar
MAHAGENCO
Mumbai, India

Nasrin I. Shaikh
Department of Chemistry
Nowrosjee Wadia College
Pune, India

Akshitha Shankar
School of Electrical Engineering
Vellore Institute of Technology
Chennai, India

Shivangi Shukla
School of Electrical Engineering
Vellore Institute of Technology
Chennai, India

Rabindra Kumar Singh
School of Computer Science and
  Engineering
Vellore Institute of Technology,
Chennai, India

M Sivabalakrishnan
School of Computer Science and
  Engineering
Vellore Institute of Technology
Chennai, India

Supraja Sivaviji
School of Electrical Engineering
Vellore Institute of Technology
Chennai, India

Srimathi R
School of Electrical Engineering
Vellore Institute of Technology
Chennai, India

N. Srinivasan
Engineering Design Research Center
L&T Construction
Mumbai, India

Srirevathi B
School of Electrical Engineering
Vellore Institute of Technology
Chennai, India

P. Srividya
Department of ECE
RVCE
Bengaluru, India

M. Subashini
School of Electrical Engineering
Vellore Institute of Technology
Chennai, India

R. Subramani
Department of Mathematics
Amrita School of Engineering
Bangalore, India

D Suganthi
School of Electrical Engineering
Vellore Institute of Technology
Chennai, India

V. Sumathi
School of Electrical Engineering
Vellore Institute of Technology
Chennai, India

S. Umadevi
School of Electronics Engineering
Vellore Institute of Technology
Chennai, India

C. Vijayalakshmi
Department of Statistics and Applied
    Mathematics
Central University of Tamil Nadu
Thiruvarur, India

V. Vijeya Kaveri
Department of Computer Science and
    Engineering
Sri Krishnan Engineering College
Chennai, India

Vinodharani M
KPIT
Bangalore, India

1 Nonlinear Controller for Electric Vehicles in Smart Buildings

Kanimozhi G and O.V. Gnana Swathika
Vellore Institute of Technology

Xiao-Zhi Gao
University of Eastern Finland

CONTENTS

1.1 INTRODUCTION

Many electric car charging solutions today are as easy as a car-to-grid link, with nothing else in between to tell either building managers or grid operators who charge when and how much electricity. But as electric cars become more popular, the need for building managers to be able to communicate with AC car chargers will grow as companies and residents try to juggle energy demands. Plug-in electric vehicle (EV) power conditioning system consists of an AC/DC boost rectifier accompanied by an independent DC/DC converter to charge a high-voltage battery bank. The front-end converter has PFC controller stage and the charger stage; the PFC stage is to enhance the input current quality received from the power supply and the charger stage is designed to energize the battery bank. Figure 1.1 shows the proposed charger for an EV. For the front-end PFC stage, boost converters with interleaving concept are generally used. In this paper, bridgeless interleaved (BLIL) boost AC/DC [1–4] converter is preferred. The main advantages of BLIL boost converter over conventional interleaved topology are increased converter's efficiency, increased reliability of the system by device paralleling, increased output voltage

DOI: 10.1201/9781003240853-1

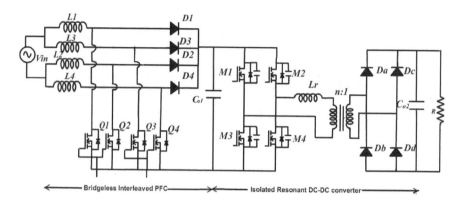

FIGURE 1.1   Bridgeless AC/DC boost converter with the isolated resonant DC/DC converter.

ripple frequency, decreased voltage and current stresses in the power semiconductor devices, decreased input current ripple, reduced passive components, and EMI filter size. The current ripple at the input side is reduced by interleaving [5], and to ensure it, the gating pulses to the switches are given with a phase shift of 180°. Here, four-channel interleaving inductors are used to reduce the input current ripple, inductor size, and output current ripple.

The popular full bridge topology [6] is chosen as the second-stage DC/DC conversion because it is highly efficient with high power density and reliability. The isolated second stage is based on resonant and PWM topology. In the traditional resonant converters, to achieve zero voltage switching (ZVS) and zero current switching (ZCS) for the power semiconductor devices, a large value of reactive current distribution for wider load variation is required. This results in bulky resonant tank and low power density. To make this possible, auxiliary circuits for commutation have been reported in the literature [7–11]. The major drawbacks of the conventional resonant converters are as follows: (i) ZVS for the lagging leg switches that are restricted for large load variations, (ii) immoderate conduction losses in passive intervals due to the circulating currents through the output inductor, and (iii) critical voltage overshoot and oscillation for diodes when the converter is operated at high voltage. Moreover, the numbers of components used in the traditional DC/DC converter are more compared to those of the proposed converter. The DC/DC converter [10], ZVS, and ZCS for lower-group power devices are attained by transformer leakage inductance, and ZCS is accomplished to turn-ON upper-group power devices.

The PFC stage and the isolated DC/DC converter stage with the conventional control arrangement are revealed in Figure 1.2. In the conventional average current mode (ACM) controller, the converter modifies the input current and regulates the intermediate DC bus voltage. Since the converter operates for a shorter period of time at full load, the controller becomes inefficient at low or light load condition. The outer voltage loop of the ACM control scheme [12–14] has very low bandwidth and has the second harmonic ripple at intermediate DC bus voltage. This leads to sluggish response and higher overshoots during transients. The control technique is improved by enhancing the bandwidth of the outer voltage loop by adding it with ripple term of the reference output voltage. Other attempts are also done by employing digital notch

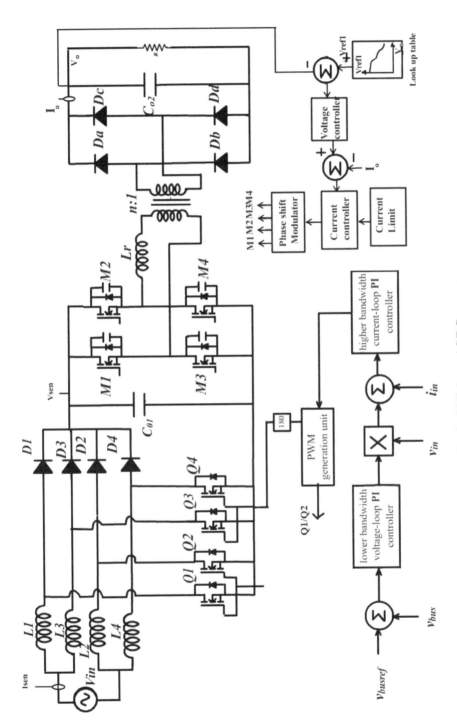

FIGURE 1.2 Conventional ACM controller implementation for BLIL boost AC/DC converter.

filter tuned at DC bus voltage to suppress the second harmonic ripple. In paper [15], to eliminate the second harmonic ripple from the DC voltage, and to enhance the converter's dynamic behavior, a dead zone digital controller was proposed.

As the front-end PFC boost rectifier has a correlation between the state variables and the duty ratio, the system's large signal model is nonlinear with wide load variation. Therefore, for a large operating state, the linear controllers cannot function optimally. Hence, the converter can be operated with high efficiency at light load conditions as the bus voltage is not fixed. In this work, the flatness controller is suggested, which works on the front-end converter's input power demand. It controls the input power instead of DC bus voltage. Thus, it eliminates the sluggish outer voltage loop and the controller with larger bandwidth to hold rapid load transient can be designed.

This paper is presented in the following ways: Section 1.2 describes the analysis and dynamic modeling of the converter, and Section 1.3 presents the implementation of ACM controller. Section 1.4 deals with the converter control using flatness theory. Section 1.5 addresses the design consideration. Section 1.6 describes the results of the simulation. Section 1.7 presents conclusion.

1.2 CONVERTER ANALYSIS AND DYNAMIC MODELING

The overall battery charger circuit, as shown in Figure 1.1, indicates BLIL PFC boost converter accompanied by full bridge isolated DC/DC converter. BLIL boost rectifier is shown in Figure 1.1, which includes four inductors (L1–L4), four (Q_1–Q_4) switches, and four (D_1 and D_4) diodes. The gating pulses to Q1 and Q2 are 180° phase-shifted with Q3 and Q4, which increases the input current ripple frequency and thus reduces the ripple. The galvanic isolation is provided by the second-stage converter using a high-frequency transformer. In this work, a phase-shifted full bridge topology with capacitive output filter has been used, which provides ZVS for the inverter switches. The lower-group devices (M3 and M4) are activated at 50% duty cycle, while the upper-group devices (M1 and M2) are PWM-controlled. The converter is operated in a discontinuous conduction mode. The upper-group switches achieve ZCS turn-on, whereas the lower-group switches achieve ZVS turn-on and ZCS turn-off.

The working principle of BLIL boost rectifier is analyzed for both positive and negative half cycle of the input AC voltage and reported in Ref. [3].

The steady-state analysis of BLIL boost converter is elucidated in Refs. [1–3]. The converter's state equations operating under continuous conduction mode are given as follows:

$$\frac{di_{L1}}{dt} = \frac{1}{2L} v_{in} - (1-D).\frac{V_{bus}}{2L} - \frac{i_{L1}}{2L} R \tag{1.1}$$

$$\frac{di_{L2}}{dt} = \frac{1}{2L} v_{in} - (1-D).\frac{V_{bus}}{2L} - \frac{i_{L2}}{2L} R \tag{1.2}$$

$$\frac{dV_{bus}}{dt} = \frac{D\left(i_{L2} - i_{L1}\right)}{C_{01}} + \frac{i_{L1}}{C_{01}} - \frac{V_{bus}}{RC_{01}} \tag{1.3}$$

where $L1 = L2 = L3 = L4 = L$ are the boost inductors, R is the load resistance, C_{01} is the intermediate capacitor bus, and i_{L1}, i_{L2} are the current entering via inductors $L1$ and $L2$. The transfer function for the second-stage resonant DC/DC converter [10] is expressed as:

$$\frac{v_O(s)}{\gamma(s)} = \frac{V_{bus}Ns\Big/Np\pi}{1 + Lr\,/\,ro.s + C_{02}s^2} \tag{1.4}$$

where $v_O(s)$ is the output voltage of the converter, $\gamma(s)$ is the phase shift between lagging and leading leg switches, Np and Ns are the number of turns in the transformer's primary and secondary, and ro and C_{02} are the incremental resistance of the load and output capacitance.

1.3 PFC BASED ON CONVENTIONAL ACM CONTROL

Figure 1.2 shows the block diagram for conventional ACM controller [12–14]. It comprised primarily the outer voltage loop and the inner current loop. The outer loop regulates the intermediate DC bus voltage, and the inner regulator shapes the input current to track the input voltage. The bandwidth of the outer voltage loop is very low, and the power factor at the input terminal is not influenced by the second harmonic ripple modulation of the capacitor bus. Now the output bus voltage v_{bus} is sensed and compared with the voltage reference v_{busref} value, and for the inner current loop, the resulting error voltage becomes the reference. The frequency 10 Hz is kept as the typical cutoff value of the voltage loop, which removes the second harmonic ripple injected in the DC bus. This lower bandwidth of the voltage loop gives a sluggish response to load variation and introduces overshoots or undershoots under transient conditions.

1.4 DIFFERENTIAL FLATNESS-BASED CONTROLLER FOR AC/DC CONVERTER

The linear controller doesn't enhance the converter's performance at a wide operating range, whereas nonlinear control technique [16] optimizes the converter performance under variable load condition. This paper talks about a nonlinear control technique for the front-end converter where the system variables are made flat with respect to the input power. Flatness principle is the measure of nonlinear behavior of the system. BLIL boost AC/DC converter is a highly nonlinear system. Flatness concept [17–21], as shown in Figure 1.3, is applied to the converter in such a way that its input current traces the sinusoidal voltage waveform to improve the power factor and controls the DC bus voltage to the desired value. The key benefit of implementing the flatness principle to the battery charger is that the design of the nonlinear system is well exemplified with respect to the flat output; therefore, by using flatness rule, the system can be converted to a basic system. The design of the controller begins with transforming the model to flat coordinates, and then with reference to such flat parameters, the controller is derived.

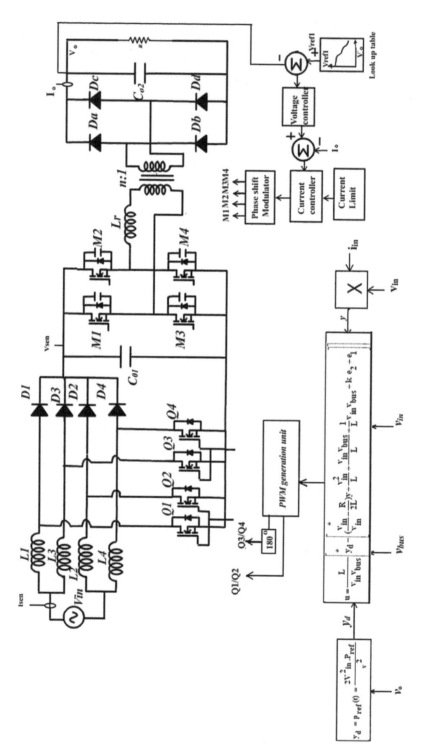

FIGURE 1.3 BLIL boost AC/DC converter with differential flatness controller.

By obtaining the set of state variables in terms of flat outputs, a system can be transformed to flat coordinates so that the model is algebraic over the differential field produced by the flat output.

The converter dynamics should be, in general, given as:

$$\overset{*}{x} = f(x,u) \tag{1.5}$$

where x and u are designated as the state variable and control variable, respectively. Then, the sets of output can be given by:

$$y = h(x,u,\overset{*}{u}.......\overset{(\Gamma)}{u}) \tag{1.6}$$

Now the inputs and the state variable are stated in terms of flat output as follows:

$$x = \varphi(y,\overset{*}{y},........,y^{(\eta)}) \tag{1.7}$$

and

$$u = \xi(y,\overset{*}{y},........,y^{(\eta)}) \tag{1.8}$$

The power reference P_{ref} for the front-end rectifier is expressed as:

$$p_{ref}(t) = P_{ref}[1 - \cos 2\omega t] \tag{1.9}$$

From equation (1.9), it is observed that the input power value varies twice the supply frequency, and hence if the input power is referred as the control variable, the outer voltage loop with low-frequency filter is not required. The instantaneous input power reference value y_d is given by:

$$y_d = p_{ref}(t) = \frac{2V^2_{in} \cdot P_{ref}}{v^2} \tag{1.10}$$

where V_{in} is the input voltage instantaneous value with v as its amplitude, and it is given as:

$$\left|V_{in}\right| = v.\sin \omega t \tag{1.11}$$

The input power p_{in} for BLIL converter is given as:

$$p_{in} = v_{in} * i_{in} \tag{1.12}$$

where i_{in} is the input current and p_{in} is expressed in flat output variable y as:

$$p_{in} \tag{1.13}$$

The variables x and u are given by equations (1.13) and (1.14) as:

$$x = i_{in};\tag{1.14}$$

$$u = D\tag{1.15}$$

where D is the duty cycle. If the system is to be flat, x and u must be expressed as a function of flat variable y, and hence, the equation for state variable x is given as:

$$x = \frac{y}{v_{in}}\tag{1.16}$$

And the control input as a flat output function is derived as:

$$\frac{dy}{dt} = x.\overset{*}{v}_{in} + v_{in}.\overset{*}{x}\tag{1.17}$$

$$\frac{\overset{*}{v}_{in}}{v_{in}}.y + v_{in}.\left[\frac{2v_{in}}{L} - \frac{(1-u)V_{bus}}{L} - \frac{R}{2L}.y\right]\tag{1.18}$$

From equation (1.17), u is derived as:

$$u = \xi(y,\overset{*}{y}) = 1 + \frac{L\overset{*}{y}}{v_{in}V_{bus}} - \left[\frac{L\overset{*}{v}_{in}}{v^2_{in}V_{bus}} - \frac{R}{2v_{in}V_{bus}}\right]y + \frac{v_{in}}{V_{bus}}\tag{1.19}$$

Thus, the system is now said to be flat as the input state variable and control variable are obtained in terms of flat variable. The system uncertainties like input disturbances and load disturbances should be included in the model as errors and the expressions are given as:

$$e_0 = \int\limits_0^t\int (y_d(\tau) - y(\tau))d\tau\tag{1.20}$$

$$e_1 = y_d - y\tag{1.21}$$

where e_0 and e_1 are the tracking variables. The control law is obtained from the Lyapunov function [15–18], and it is given as:

$$V(e_1,e_2) = \frac{1}{2}e_1^2 + e_2^2\tag{1.22}$$

The derivative of equation (1.22) is given by:

$$\overset{*}{V}(e_1,e_2) = e_1.e_2 + e_2\left[\left(\frac{v_{in}^{*}}{v_{in}} - \frac{R}{2L}\right)y - \frac{v_{in}^2}{L} - \frac{v_{in}v_{bus}}{L} + \frac{1}{L}v_{in}v_{bus}u\right] \quad (1.23)$$

Thus, the control law is obtained on simplifying equation (1.22) in terms of u as:

$$u = \frac{L}{v_{in}v_{bus}}\left[\overset{*}{y_d} - \left[\left(\frac{v_{in}^{*}}{v_{in}} - \frac{R}{2L}\right)y + \frac{v_{in}^2}{L} + \frac{v_{in}v_{bus}}{L} + \frac{1}{L}v_{in}v_{bus} + k\,e_2 + e_1\right]\right]$$

$$(1.24)$$

and its derivative is expressed as:

$$\overset{*}{V} = -k_1.e_2^2 \quad (1.25)$$

Thus, from the calculated instantaneous power and its reference value, the tracking error e_1 and its integral e_0 are obtained. The duty cycle for the switches is derived from equation (1.24) by using Lyapunov control law [17].

1.5 DESIGN CONSIDERATIONS

This section explains the designing of the proposed two-stage battery charger. The four-channel interleaving inductors of BLIL boost rectifier are designed with 20% as input ripple current, and it is expressed as:

$$\Delta I_{:L} = \frac{V_{inp}\sqrt{2}D}{fs.\dfrac{Lb}{2}} \quad (1.26)$$

where $V_{inp} = \sqrt{2}Vs\sin\omega t$ is the peak value of the input voltage; f_s is the switching frequency of the rectifier; $Lb = L1 = L2 = L3 = L4$ are the boost inductors. Two inductors of equal value are connected to each phase. D is the duty cycle, and it is expressed as:

$$D = 1 - \frac{V_{inp}}{V_b} \quad (1.27)$$

where V_b is the output voltage of the boost converter. The capacitor is given by the following expression:

$$C_{01} = \frac{2Po*T_h}{V_b^2 - (V_b*0.75)} \quad (1.28)$$

where P_0 is the output power and T_h is the maximum hold-up time for the line frequency 50 Hz. The output power P_0 is given as:

$$Po = V_b * I_b \tag{1.29}$$

where I_b is the rectifier output current.

The voltage gain (G) for the second-stage converter is formulated as follows:

$$G = \frac{V_o}{Vdc} = \frac{2*n}{1 + \sqrt{1 + \dfrac{4*K}{D^2}}} \tag{1.30}$$

where n is the turn's ratio of the transformer and K is the normalized time constant. The turn's ratio is expressed as (1.31).

For inverter switches, the duty ratio is set as 0.377 as it offers optimum value for gain. From equation (1.31), the turn's ratio of the transformer is obtained as 1.326. The voltage gain has been calculated and strategized using MATLAB software for various ranges of D from 0.1 to 0.5 and k from 0.1 to 1, as shown in Figure 1.4. The resonant inductor value (L_r) is expressed as:

$$L_r = \frac{k * R * T}{4 * n^2} \tag{1.32}$$

where R is the output resistance of the converter (Ω) and T is the switching period. Thus, the value of L_r from equation (1.32) is found to be 176 µH. Figure 1.4 gives the prototype's critical value.

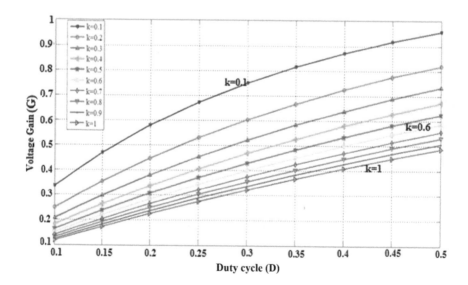

FIGURE 1.4 Plot for various k values.

1.6 CONTROLLER REALIZATION THROUGH SIMULATION

The simulation is done using PSIM 9.1.0 simulator. The circuit specifications of the bridgeless boost converter are as follows: output power $Po = 300\,\text{W}$, input voltage $(V_{in}) = 230\,\text{V}$, switching frequency $(f_s) = 70\,\text{kHz}$, inductors $(L1 = L2 = L3 = L4) = 0.58$ mH, and capacitor $(C0) = 470\,\mu\text{F}$. The transient behavior of the converter is analyzed under step load change. DSP TMS320LF28335 is chosen as it is a floating-point processor and computes complex mathematical calculation. The closed-loop simulation for BLIL is analyzed with both the control techniques. Figure 1.5 shows the simulation results of converter under digital ACM control implemented in DSP.

The input current tracks the input voltage and the output voltage is maintained at 400 V, as shown in Figure 1.5a. A positive and negative step load change is introduced at $t = 0.38$ seconds to observe the transient behavior of the system, as shown in Figure 1.5b and c. The sluggish response of the controller can be realized by the settling time of the output voltage. The steady state is attained by taking more than five cycles (i.e., $t \geq 0.1$ seconds).

The power circuit controlled by flatness theory is shown in Figure 1.6a. The input voltage and the input current waveform portrayed in Figure 1.6b reveal that the power factor is closer to unity. The positive and negative load changes at $t = 0.38$ seconds are depicted in Figure 1.6c and d, respectively. It is observed that the input power tracks its power reference value very fast, which results in regulating the output voltage with no undershoot or overshoot [21] in the DC bus voltage.

The gating pulses $vg1$ and $vg2$ with the trailing edge are given for the upper-group devices $M1$ and $M2$ with a duty cycle of 37.77%. The lower-group devices $M3$ and $M4$ are triggered with $vg3$ and $vg4$ G_4 (50% of the duty cycle). As $M1$ and $M4$ switches trigger, 400 V (peak to peak) input voltage appears across the transformer primary winding with the input current of 2A, as shown in Figure 1.7b. The voltage is null at one time instant, as shown in Figure 1.7, which states a passive interval in a discontinuous conduction mode. The secondary winding voltage is decreased to 200 V (peak to peak) and current to 1A (peak to peak) with transformer turns ratio 1.326:1, as shown in Figure 1.7c. ZVS ON and ZCS OFF are accomplished for the lower-group switches on inverter side. The switch $M3$ drain to source voltage and the resonant inductor current are displayed in Figure 1.7d. For the diodes on the rectifier side, which can be observed from Figure 1.7e, the ZCS is accomplished.

From the comparison graph revealed in Figure 1.8, it is renowned that the converter efficacy is high for both the controllers at heavy load conditions, whereas at light load, the flatness controller provides higher efficiency than the ACM controller.

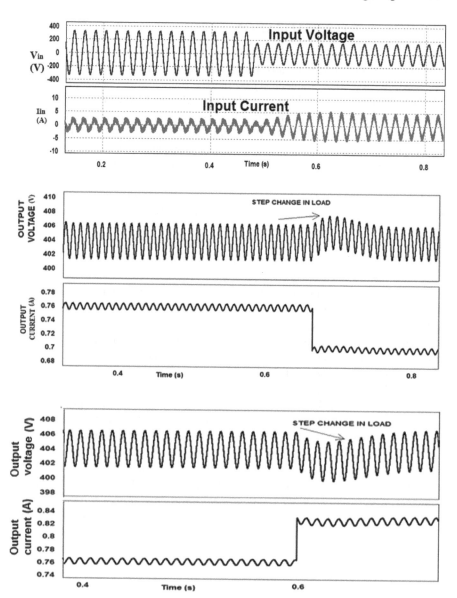

FIGURE 1.5   (a) Input voltage and waveform of input current. (b) Transient behavior of output voltage with positive step load change. (c) Output voltage transient behavior with negative step load change.

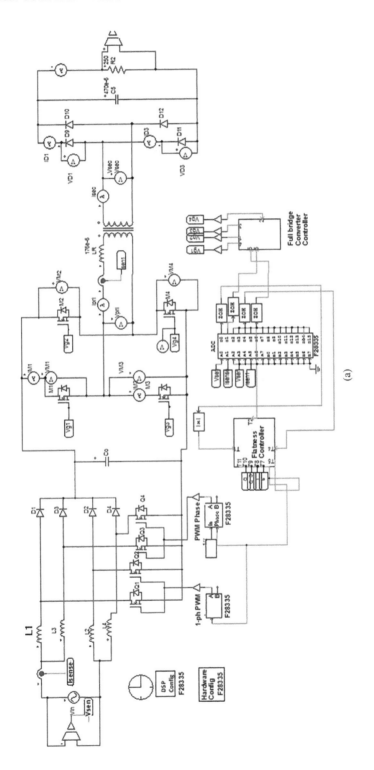

(a)

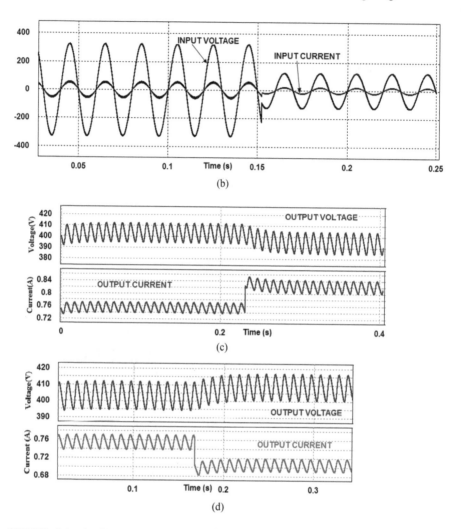

FIGURE 1.6 (a) PSIM model of flatness controller-based BLIL boost converter. (b) Waveforms of input voltage and input current. (c) Transient performances of output voltage and current with positive phase change in load. (d) Transient behavior of load voltage and the current with negative phase change in load.

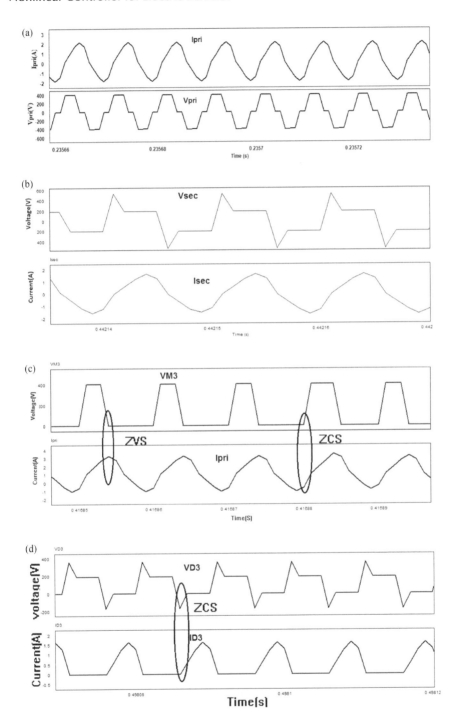

FIGURE 1.7   (a) Current and voltage across transformer input winding. (b) Voltage and current through the secondary transformer winding. (c) Zero voltage switching and zero current switching of M3. (d) Zero current switching of diode D3.

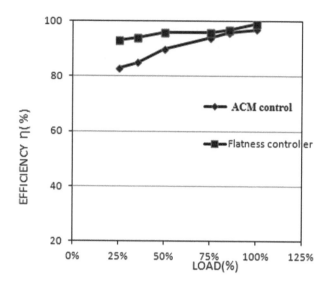

FIGURE 1.8   Efficiency vs. load variation.

1.7   CONCLUSION

This work presented a nonlinear control technique based on the flatness principle for BLIL PFC boost converter to control the input power as the EV is charged in the smart buildings. The front-end converter shows a fast and dynamic response in regulating the output voltage under load perturbations compared to the ACM control. Here, the output voltage is regulated based on the input power instead of intermediate DC bus voltage. The second-stage converter attains soft switching, which improves the overall efficiency of the converter. Thus, this flatness control shows superior performance of the converter under light load conditions in regulating the output voltage and improving the input power factor of the front-end AC/DC converter.

REFERENCES

[1] F. Musavi; W. Eberle; W.G. Dunford, "Efficiency evaluation of single-phase solutions for AC-DC PFC boost converters for plug-in-hybrid electric vehicle battery chargers," in *IEEE Vehicle Power and Propulsion Conference*, Lille, France, 2010.

[2] F. Musavi; W. Eberle; W.G. Dunford, "A phase shifted semi-bridgeless boost power factor corrected converter for plug in hybrid electric vehicle battery chargers," in *IEEE Applied Power Electronics Conference and Exposition*, APEC Fort Worth, TX, 2011.

[3] F. Musavi; W. Eberle; W.G. Dunford, "A high-performance single-phase AC-DC power factor corrected boost converter for plug in hybrid electric vehicle battery chargers," in *IEEE Energy Conversion Congress and Exposition*, Atlanta, Georgia, 2010.

[4] G. Kanimozhi; V.T. Sreedevi, "Improved resettable integrator control for a bridge-less interleaved AC/DC converter," *Turkish J. Electric. Eng. Comput. Sci.* 2017, 25, 5, 3578–3590.

[5] G. Kanimozhi; V.T. Sreedevi, "Semi-bridgeless interleaved PFC boost converter for PHEV battery chargers," *IETE J. Res.* 2018, 1–11.

[6] T.H. Kim; S.J. Lee; W. Choi, "Design and control of the phase shift full bridge converter for the on-board battery charger of electric forklifts," in *Proc. IEEE Power Electronics and ECCE Asia, ICPE and ECCE*, 2011, pp. 2709–2716.

[7] D.B. Dalal, "A 500 kHz multi-output converter with zero voltage switching," in *Proc. IEEE Appl. Power Electron. Conf. Expo.*, 1990, pp. 265–274.

[8] G.-B. Koo; G.-W. Moon; M.-J. Youn, "Analysis and design of phase shift full bridge converter with series-connected two transformers," *IEEE Trans. Power Electron.* 12, 2, 411–419, 2004.

[9] A.J. Mason; D.J. Tschirhart; P.K. Jain, "New ZVS phase shift modulated full-bridge converter topologies with adaptive energy storage for SOFC application," *IEEE Trans. Power Electron.* 23, 1, 332–342, 2008.

[10] K. Likhitha; S. Sathish Kumar; G. Kanimozhi, "Isolated DC/DC zero voltage switching converter for battery charging applications," in *2016 Biennial International Conference on Power and Energy Systems: Towards Sustainable Energy (PESTSE) at Bengaluru, India*, 2016.

[11] M. Borage; S. Tiwari; S. Bhardwaj; S. Kotaiah, "A full-bridge DC–DC converter with zero-voltage-switching over the entire conversion range," *IEEE Trans. Power Electron.* 23, 4, 1743–1750, 2008.

[12] S. Maulik, "Power factor correction by interleaved boost converter using PI controller," *Int. J. Eng. Res. Appl.* 5, 5, 918–922, 2013.

[13] K.P. Hase; K. Louganski, "Generalized average-current-mode control of single-phase ac-dc boost converters with PFC," *IEEE Trans. Power Electron.* 30, 2007.

[14] M. Veerachary; T. Senjyu; K. Uezato, "Modeling of closed-loop voltage-mode controlled interleaved dual boost converter," *Comput. Electr. Eng.* 29, 6784, 2003.

[15] A. Prodic; D. Maksimovic; R.W. Erickson, "Dead-zone digital controllers for improved dynamic response of low harmonic rectifiers," *IEEE Trans. Power Electron.* 21, 1, 173–181, 2006.

[16] M. Pahlevaninezhad; P. Das; J. Drobnik; S. Member, "A nonlinear optimal control approach based on the control-lyapunov function for an AC/DC converter used in electric vehicles," *IEEE Trans. Power Electron.* 28, 324–328, 2012.

[17] M. Pahlevaninezhad; P. Das; A. Servansing; P. Jain; A. Bakhshai; G. Moschopoulos, "An optimal Lyapunov-based control strategy for power factor correction AC/DC converters applicable to electric vehicles," *2012 Twenty-Seventh Annual IEEE Applied Power Electronics Conference and Exposition (APEC)*, Orlando, FL, 2012, pp. 324–328.

[18] S. Kumar; K.G. Gunturi; V.T. Sreedevi, "Comparative analysis of digital linear and non-linear controller for a PFC boost converter," *Int. J. Appl. Eng. Res.* 10, 20, 15689–15693, 2015.

[19] B. Lantos; L. Rinc; M. Rton, *Non-Linear Control of Vehicles and Robotics*, Springer, 2011.

[20] M. Brenna; F. Foiadelli; C. Leone; M. Longo, "Electric vehicles charging technology review and optimal size estimation," *J. Electr. Eng. Technol.* 15, 2539–2552, 2020. https://doi.org/10.1007/s42835-020-00547-x.

[21] S.M. Majid Pahlevaninezhad, "A new control approach based on the differential flatness theory for an AC/DC converter used in electric vehicles," *IEEE Trans. Power Electron.* 27, 4, 2085–2103, 2012.

2 Guidance System for Smart Building Using Li-Fi

Rabindra Kumar Singh, B V A N S S Prabhakar Rao, and M Sivabalakrishnan
Vellore Institute of Technology

CONTENTS

2.1 INTRODUCTION

India is a large country, and around the world, India is the second-most populous country and has fast-growing economy. We depend on navigation systems to help us in our everyday life activities. The evolution of GPS has brought a major change to the navigation system around the world. Everyone uses a navigation system in order to do daily activities. All these systems are maintained and serviced by satellite systems. Many universities, MNCs, and shopping malls around the world have multistoried buildings that cover a large area. For a new person, going to a particular location in those areas is difficult. He has to either ask people for help or should look for signboards for guidance.

DOI: 10.1201/9781003240853-2

Even if the person decides to use GPS for navigation purposes, it would not serve the purpose here as the GPS changes within a building are very small and we cannot achieve that accuracy using GPS alone. Also, when we speak about multiple floors, GPS can't guide users in this aspect. So, in order to overcome all these drawbacks, we have come up with a new system, which can serve this purpose. Many malls and universities these days have signboards and maps of the location at the entrance, but for a person to remember the map and follow the route is not optimal. So we will be using the existing systems to come up with a solution to this problem. Everyone uses smartphones these days – this is one of the points that can be exploited in designing our guidance system.

Now smart buildings are in reality that exist around us. Smart buildings are getting new shape and architecture as the technology keeps growing. Several researchers have emphasized intelligent and smart building in their own way [1–4]. The guidance system is an important application, which plays a major role in day-to-day life. Finding the correct location in a complex building such as universities or shopping malls is always a challenging task. Consider a smart building where in everything is to be smart then why not navigation system as well. Finding the location of the users and guiding them to their destination are the two major functionalities of a guidance system. The main idea of this chapter is to create internal navigation systems for a confined space to create automatic navigation for the visitors using Li-Fi (Light-Fidelity) technology. The existing systems use sign boards or announcement systems for internal navigation. There are few places where we can even see map of entire place in the entrance. But all these systems are rigid and can't fully help in guidance. Li-Fi is involved in the transmission of data using visible light by sending data through a LED light bulb that varies in intensity faster than the human eye can follow. The term "Li-Fi" refers to visible light communication (VLC) technology that uses light as a medium to deliver high-speed communication in a manner similar to Wi-Fi.

2.2 LITERATURE REVIEW

2.2.1 TRANSFERRING DATA USING LI-FI TECHNOLOGY

GPS seems to be the best solution to develop outdoor location systems, but performance of these systems is not good enough to locate entities within indoor environments, mainly if accuracy and precision are required. In this chapter, we propose a tracking indoor system based on Li-Fi technology, which does the part of navigation and location specification [5–10].

2.2.2 CONSTRAINED AND QUANTIZED GRAPH ALGORITHM
TO FIND THE SHORTEST PATH

All the destinations and master consoles are considered as nodes. All the paths between all nodes are fixed. The problem that is addressed here is the floor-based navigation. Each floor is considered as a subgraph, and the nodes in that floor constitute for that subgraph. "n" number of subgraphs represent the "n" floors in the building. All the "n" subgraphs together make the main graph. The shortest-path algorithm has to be applied on this main graph [11–15].

2.2.3 Bluetooth Technology for Communication

Bluetooth is easily accessible and easily available technology. All the mobile phones support Bluetooth. Since our idea is to create an android app for users to use the navigation system, using Bluetooth will help us cut the boundaries, which may act as constrains [16–20].

2.3 EXISTING SYSTEM

GPS: Even though GPS solves most of navigation issues, a few special cases like internal navigation pose a problem. Although GPS is more sophisticated than Li-Fi navigation, the lack of options forces us to swing towards Li-Fi. Minor changes cannot be effectively noted and users can't be guided effectively.

2.4 PROPOSED SYSTEM

A guidance system for smart building using Li-Fi technology has been proposed in this part of the chapter. The system has been divided into several modules, in which the advantages of the proposed system are emphasized.

The Li-Fi technology is used in the proposed system to provide location-based positioning and also used to communicate the details to the user. This can help people to navigate without any external help with ease. The proposed system coins a low-cost guidance system that amalgamates the potency of Li-Fi device.

The entire workflow can be explained better using a flowchart. This explains the overall flow of the system. The system can be mainly divided into two parts, which are explained and discussed below.

Figure 2.1 represents the flow of the entire system. The main elements of the system are all listed here. The entire system starts with users' input through the

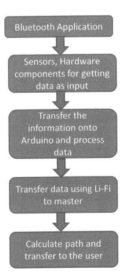

FIGURE 2.1   Flow of system.

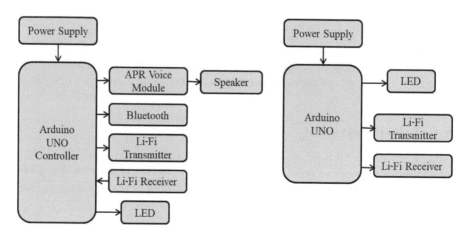

FIGURE 2.2   System architecture.

mobile and ends with system returning the request back to user. Many back-ground processes that happen are hidden in this overview. We will be discussing in deep about the modules present and how they all work together to get the final output.

2.5   ARCHITECTURE OF THE SYSTEM

Figure 2.2 represents the entire architecture of the system. The RX section is oth-erwise called sub-console of the project. TX section is called the main console. The RX section is used by the user to connect to the main console and get required details.

The image on the left is sub-console, which communicates with the user and delivers their requirements to the main console. Image on the right is the main con-sole, and it processes the user requests and sends appropriate response to sub-console. Later, the sub-console relays the data over to the user.

2.6   WORKFLOW

Figure 2.3 is the module 1 of the system, which is also called as sub-console. This module acts as bridge between the user and the main console. This gets data from the user and transfers it to the main console to get the path. It also receives the path from the main console and transfers it to the user. Connection establishment and data transfer occur in this module.

In Figure 2.4, we have module 2, which takes care of applying the shortest-path techniques and giving the details. It receives the data from the sub-console that applies the shortest-path algorithm and gets optimum path, which should be taken in order to reach the destination. The entire system is divided into several modules, which have been explained in detail.

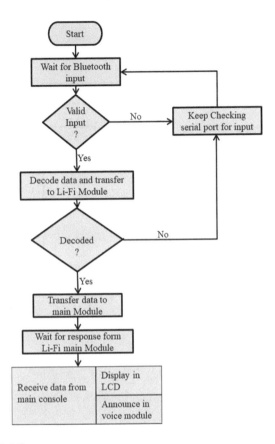

FIGURE 2.3   Workflow.

2.7   MODULES

2.7.1   MODULE 1

This module consists of an Android app, which will be the source of communication between the user and the Li-Fi guidance system. The destination details are communicated using app by the user. The user first connects to the Bluetooth module (HC-05) present in the slave module of the system. The user then sends the destination details through the app to the module.

These details are sent over to the Arduino where data is decoded. The Bluetooth always sends data as single bits of Ascii values. These values are decoded to get the destination details.

2.7.2   MODULE 2

This module consists of the slave Arduino. This decodes the data sent over by the Bluetooth module. As the data is received in Ascii, we convert into normal data and send the data over to the master console. All the routes are calculated only on the

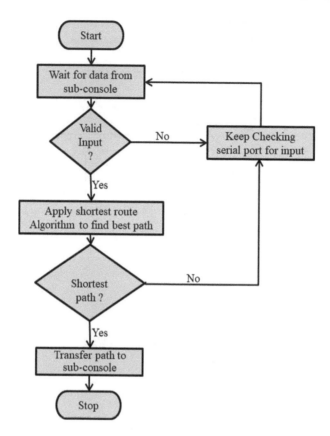

FIGURE 2.4   Shortest-path flowchart.

master side. The received data is sent using Li-Fi transmitter. Once the data is sent successfully, it then waits for the data from the master console. The master console sends the path to the slave, and it is received using Li-Fi receiver. This data is later displayed on LCD display.

2.7.3   MODULE 3

This module consists of the master console, which handles the data received from slave console and analyzes it. Once the destination address is received, it then calculates the shortest path to reach the destination using Dijkstra's algorithm and it sends the results over to the slave control.

2.7.4   MODULE 4

As soon as signal is received from module 2, the main module knows the present location and the paths to all the locations in the locality. So here we implement a shortest-path algorithm using subgraphs. Once we get to know the shortest path, the details are communicated to the slave module.

2.7.5 ALGORITHM

1. Make a set that tracks all vertices in the graph; this set is empty in the beginning of the algorithm.
2. Allocate a distance value to all vertices. For the starting vertex, the value is set to 0 and the rest of all are infinity.
3. The present available set doesn't have all vertices.
4. Pick a new vertex, which is nearest to the starting node and which is not present in the set.
5. Include this new vertex into the set.
6. Update adjacent vertices distance values.
7. To do this, visit all adjacent vertices of the present vertex(x), and if distance of x from the initial vertex and the value of edge between x and adjacent vertex is lesser, then update the adjacent vertex value.

In Figure 2.5, there are four different blocks. Each block represents a different floor, and each floor has all the nodes in it connected and reachable. But between different floors, we can only have one path and its weight is different from other paths. We can use Dijkstra's algorithm on this prepared graph and can get the optimum path to the destination.

If 3 is the source node and 15 is the destination node, as we have seen in the previous algorithm, it gives us the exact steps taken to reach the destination. So, if we consider this graph, the path would be 3-0-4-7-13-15.

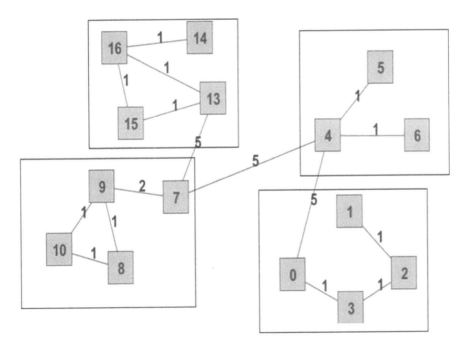

FIGURE 2.5   Path from origin to destination.

2.8   CONSTRAINTS

1. Transmission: The main catch here is the transmission range, which is only where the LED bulbs transmit light.
2. Garbage values: Small deviations from the position may lead to receiving garbage data, which can't be processed correctly.
3. Environmental constraints: The external forces also influence transmission; if the path of data transmission is blocked or if there are disturbances in receiving data, then also we lose precious data.
4. Bluetooth connectivity: Since the model is based on Bluetooth, the user should be in range of Bluetooth and should be connected in order to relay the data.
5. No data: There are cases when there is no reading from the device, and this time the solution is to restart the system.
6. Security issues: One of the major constraints of the model is the security issues regarding the priority mode. If multiple requests come exactly at the same time since the initial transfer occurs through Bluetooth, there is a chance that one of requests might not get a response.

2.9   CONCLUSION

An effective solution is proposed in this paper by considering an approach, which can help in internal navigation. This paper proposes the combination of new emerging technology Li-Fi and combines it with graph theory concepts to achieve efficient results. The main aim is to come up with a solution for internal navigation using Li-Fi. Our system provides guidance in a confined space. The use of optimizing algorithms helps us to deliver the best path available to the users from their current position. Also, the use of Bluetooth and Li-Fi technologies helps us in reducing the product cost by large extent. Also, the use of Li-Fi is a major positive point of the project. Li-Fi is new emerging technology, which can be applied in many other fields. This implementation can become ground for many future ideas, and also, there is huge scope of development.

2.10   FUTURE WORK

The project can be extended to make it practical by adding a new application, which can make the user have easier access to the navigation system. Also, the Li-Fi technology can be enhanced further to increase the data streaming quality, which can help to reduce garbage values in the transfer. This project can be enhanced in a few ways to make it more user-friendly. A few changes can be made, which can make the project to improve its functionality. The Android app can be improvised to give responsive feedbacks to the user. A local map can be added to the app, using which the user can view the entire locality and can select the destination he wants to go.

REFERENCES

[1] Froufe, M.M.; Chinelli, C.K.; Guedes, A.L.A.; Haddad, A.N.; Hammad, A.W.A.; Soares, C.A.P. Smart buildings: Systems and drivers. *Buildings* 2020, 10, 153. https://doi.org/10.3390/buildings1009015.

[2] Lima, E.G.; Chinelli, C.K.; Guedes, A.L.A.; Vazquez, E.G.; Hammad, A.W.A.; Haddad, A.N.; Soares, C.A.P. Smart and sustainable cities: The main guidelines of city statute for increasing the intelligence of Brazilian cities. *Sustainability* 2020, 12, 1025.

[3] Marikyan, D.; Papagiannidis, S.; Alamanos, E. A systematic review of the smart home literature: A user perspective. *Technol. Forecast. Soc. Chang.* 2019, 138, 139–154.

[4] Rao, B.P.; Singh R.K., Disruptive intelligent system in engineering education for sustainable development, *Procedia Comput. Sci.* 2020, 172, 1059–1065.

[5] Simon, P.G., Robust and accurate indoor localization using visible light communication. *IEEE* 2017.

[6] Pindoriya, N.M.; Dasgupta, D.; Srinivasan, D.; Carvalho, M., Infrastructure security for smart electric grids: A survey, In *Optimization and Security Challenges in Smart Power Grids (Energy Systems)*, Berlin, Germany: Springer, 2013, pp. 161–180.

[7] Guidelines for smart grid cybersecurity: Supportive analysis and references, NISTIR, Tech Rep. 7628, Aug. 2010. http:csrc.nist.gov/publications/nistir/ir7628/nistir-7628-vol2.pdf.

[8] Haruyama, S., "Visible light communication using sustainable LED lights," in *2013 Proceedings of ITU Kaleidoscope: Building Sustainable Communities*, Kyoto, 2013, pp. 1–6.

[9] Deng, P.; Kavehrad, M.; Kashani, M.A., "Nonlinear modulation characteristics of white LEDs in visible light communications," in *Optical Fiber Communication Conference, OSA Technical Digest (online) (Opti-cal Society of America, 2015)*, paper W2A.64.

[10] Antony, J.; Verma, P., Exploration and supremacy of Li-Fi over Wi-Fi. *Int. J. Comput. Appl. Technol. Res.* 2016, 5, 2, 83–87. ISSN:-23198656 www.ijcat.com.

[11] Orr, R.; Abowd, G., "The smart floor: A mechanism for natural user identification and tracking," in *Proceedings of the 2000 Conference on Human Factors in Computing Systems (CHI '00)*, 2000.

[12] Ghansah, I., "Smart grid cyber security potential threats, vulnerabilities and risks," in *Public Interest Energy Res. (PIER) Prog. Interim Rep., California State Univ., Sacramento, Sacramento, CA, USA, Tech. Rep. CEC-500–2012–047*, 2012, pp. 73–83.

[13] Azari, A., Survey of smart grid from power and communication aspects. *Middle East J. Sci. Res.* 2014, 21, 9, 2014.

[14] Usha, S.; Sangeetha, P., "Multiple attribute authority based access control and anonymous authentication in decentralized cloud," 2016, pp. 24–29.

[15] Chan, C.F.; Zhou, J., CyberPhysical device authentication for the smart grid electric vehicle ecosystem. *IEEE J. Select. Areas Commun.* 2014, 32, 7, 1509–1517.

[16] Asadullah, M.; Ullah, K., "Smart home automation system using Bluetooth technology," in *2017 International Conference on Innovations in Electrical Engineering and Computational Technologies (ICIEECT)*, IEEE, 2017, pp. 1–6.

[17] Baronti, P.; Pillai, P.; Chook, V.W.; Chessa, S.; Gotta, A.; Hu, Y.F., Wireless sensor networks: A survey on the state of the art and the 802.15. *Comput. Commun.* 2007, 30, 7, 1655–1695.

[18] Ramlee, R.; Leong, M.H.; Singh, R.; Ismail, M.M.; Othman, M.A.; Sulaiman, H.A.; Misran, M.H.; Said, M.A.M., Bluetooth remote home automation system using android application. *Int. J. Eng. Sci.* 2013, 2, 1, 149–153.

[19] Saxena, N.; Choi, B.J., State of the art authentication, access control, and secure integration in smart grid. *Energies* 2015, 8, 11883–11915.

[20] Wan, W.; Lu, Z., Survey cyber security in the smart grid: survey and challenges. *J. Comput. Networks: Int. J. Comput. Telecommun. Network. Arch.* 2013, 57, 5, 1344–1371.

3 Smart Building Automation System

P. Srividya
RVCE

CONTENTS

3.1 INTRODUCTION

Over a last few decades, there is a change in the working pattern in various companies. This demands for an accommodation that suits the requirements. Most of the companies look out for buildings that attract best workmen and inspire them to perform better, support good business by attracting potential customers, reduce the costs of building operations, and enhance the comfort and performance. These requirements are met by making the building smart by incorporating technologies like Internet of Things (IoT), Big Data, and Artificial Intelligence (AI). Smart building transforms the work, the workplace, and the landscape by offering substantial benefits over conventional one for both owners and occupants.

Incorporating intelligence in building is achieved by deploying sensors and actuators throughout the building and integrating them with systems like building management system (BMS), building automation system (BAS), and others, as shown in Figure 3.1.

The smart building system is not an independent unit. It is rather a system of systems as shown in Figure 3.2.

DOI: 10.1201/9781003240853-3

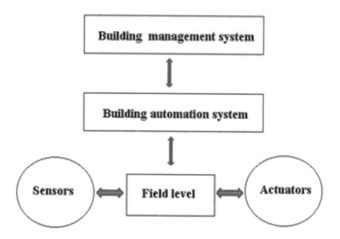

FIGURE 3.1   Smart building system.

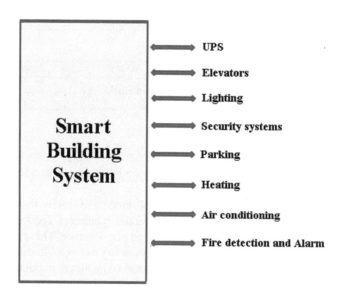

FIGURE 3.2   Smart building as system of systems.

3.2   SMART BUILDING ARCHITECTURE

Smart building architecture is divided into different levels [1], as shown in Figure 3.3.

- **Physical level**: It has interface services to different building systems like heating, ventilating, and air conditioning system (HVAC), lighting, fire, security etc. The interface services include adapters and connectors. At this level, sensors collect the data and actuators perform the required task. For example, lighting level can be sensed and adjusted automatically to match

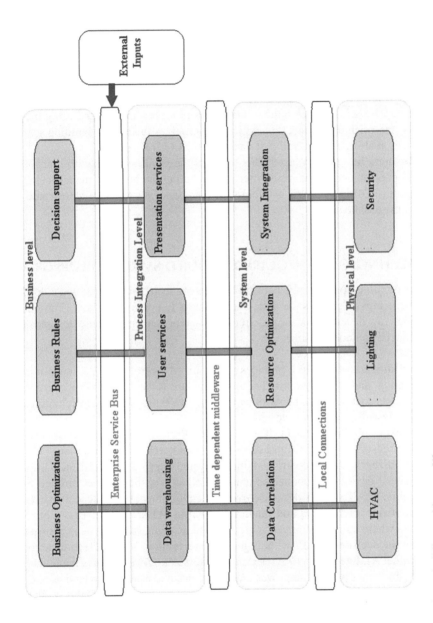

FIGURE 3.3 Smart building architecture [1].

the time of the day. The systems in this level communicate with each other through common information and operational model. The adapters are linked through local connections and use industry standard protocols.

- **System level**: It provides physical-level integration of different building systems. This level interacts with adapters in physical level through local connections. The information is in the form of signals and commands between the subsystems. Some of the services provided by this level include data archiving, event correlation, and others.
- **Process integration level**: This level provides automation and process synchronization. The disparate software services are connected by an enterprise service. The build management processes are created using the service-oriented architecture, which synchronizes the events, building services at abstract level, and past data such that it is reusable.
- **Business integration level**: This provides the business users with graphical presentation of the result after implementing the building management process. Offline analysis and real-time analysis of the data are represented on the dashboard.

3.3 TECHNOLOGIES REQUIRED TO BUILD SMART BUILDINGS

A smart building consists of a network of interconnected sensors to generate raw data, microchips to process the data, and actuators to control the data. Conversion of complex raw data to tailor the requirement is a real challenge. These demands for few advanced technologies like:

1. **Building automation**: It provides a centralized control of HVAC, security, lighting, and other systems. These systems are IoT compatible and have a capacity to monitor and control parameters like temperature, water, humidity, pressure, electricity, etc.
2. **Internet of Things (IoT)**: It uses Internet Protocol (IP) platform to connect sensors and microchips and to analyze and exchange information. This optimizes the device control automatically and improves the efficiency. IoT-enabled devices are connected and controlled over the internet through a mobile app. This provides a user-friendly environment for those who use it.
3. **Artificial intelligence and machine learning (AI and ML)**: AI is a process where human intelligence is simulated using machines. The machines are the decision makers without human intervention. ML is a technique to realize AI. Integration of AI with IoT sensors not only automates the building functions but also improves the operations, makes it more adaptive, and decreases the inefficiencies. This ensures more structural reliability and reduced ecological impacts.
4. **Augmented reality (AR)**: It provides a modified view of the actual environment. The elements are accompanied by system-generated sensory inputs. It involves a camera and some sort of viewing devices like smartphone or eyeglasses. AR superimposes a non-real object into the physical environment.

This is helpful in construction sites. AR allows blueprints to be viewed in 3D, even allows viewing the upcoming stages of construction, and helps to create extraordinary precision and accuracy.

5. **Virtual reality (VR)**: It helps the architects to experience the walk around the building much before they are built. This provides them to redesign the structure precisely based on the experience.

3.4 BUILDING AUTOMATION SYSTEM (BAS)

The main aim behind the smart building automation project is to obtain a reduction in energy consumption by efficient energy utilization and greenhouse gas emission. Energy consumption of buildings can be reduced by using energy-efficient light bulbs and efficient insulators [2]. Further reduction can be obtained by automation and control integrated with various advanced technologies. The services provided by smart buildings offer a value-added functionality and help in easier operation. Both the owners and residents relish the benefits of comfortable living and monetary gain.

A smart building is a fully automated structure with connected electronic appliances over a personal area network. For this operation, Wi-Fi facility is required throughout the day. This allows all the gadgets accessible from any corner of the world.

BAS is based on the sensors and actuators placement throughout the building and their interconnection over the network. The general Smart Building Automation System (SBAS) structure is as shown in Figure 3.4.

Sensors are required to monitor the environmental variations continuously both inside and outside the building. The various environmental variations include temperature, pressure, humidity, and lighting. The sensor data is fed to actuators to control the devices. The architecture includes both hardware and software.

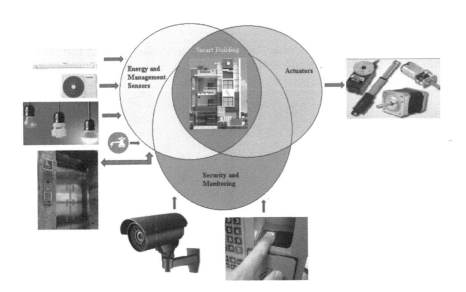

FIGURE 3.4 Smart building automation system.

Hardware includes:

- A processor unit, memory unit (flash memory for application and OS storage, SDRAM for application and OS program execution, SRAM for data storage)
- Suitable analog-to-digital converter (ADC) and digital-to-analog converter (DAC)
- High-resolution PWM modules
- Variable-frequency drive for controlling motors
- Signal conditioning ICs
- Sensors
- Relays
- Actuators
- Zigbee wireless modules
- Interfaces for gateways
- Ethernet interfaces.

Software includes:

- Diagnostic software for interfaces
- Software to program the processor unit.

3.4.1 Heating, Ventilating, and Air Conditioning System (HVAC System)

HVAV system is a collection of components that work together to deliver heat, to ventilate a space, and to remove heat from the required area. In the existing system, these units exist independently and do not react according to the internal and external climate and lead to energy wastage. This is overcome by using a smart HVAC system.

3.4.2 Sensors

Sensors are the devices that detect and respond to optical or electrical signals. They convert the physical parameter into a signal that can be electrically measured. Various sensors required to build smart building are:

1. Temperature sensor
2. Humidity sensor
3. Motion sensor
4. Gas/air-quality sensor
5. Electrical current-monitoring sensor
6. Ultrasonic sensor.

Temperature sensor: It measures the amount of heat in the room to sense the temperature change. It is used to control heating and air conditioning of the room. There exist four types of temperature sensors:

a. **Semiconductor-based sensor**: It is compact, produces linear outputs, has a small temperature range, and is of low cost.

b. **Thermocouple**: It consists of two wires made out of different metals. These wires are soldered together at one end, forming a junction. The temperature is measured at this junction. The change in temperature at the junction creates a voltage.

c. **Resistance temperature detector (RTD)**: The temperature is determined by measuring the resistance of a wire. Usually, copper, nickel, or platinum wires are used.

d. **Negative temperature coefficient thermistor (NTC thermistor)**:It is a resistive temperature sensor. The resistance decreases with a rise in temperature.

Humidity sensor: It measures the humidity in the environment and converts it into corresponding electrical signal. Relative humidity is a measure of the amount of water vapor in the atmosphere. Too much of moisture content causes condensation and results in corrosion of the machines. Common types of humidity sensors are:

a. **Capacitive**: Appropriate dielectric material is placed between two electrodes to measure the relative humidity. The dielectric constant of the material changes with the relative humidity of the atmosphere.

b. **Resistive**: Operation is similar to capacitive sensor. Except that, it uses ions in salts to measure the change in resistance. The electrical impedance of atoms is measured using the ions in salts. The resistance of the electrodes placed in the salt medium changes with the humidity variation.

c. **Thermal**: Based on the humidity of the surrounding environment, two thermal sensors conduct electricity. One of the sensors is sheathed with dry nitrogen, and the other sensor measures the surrounding air. The difference between the two is the measure of humidity.

Motion sensor: It is used to sense the movement by detecting the heat released by people occupying a particular area. It alerts when someone enters the region.

Gas/air-quality sensor: It helps in monitoring the air quality and to detect the presence of various gases. This will be helpful in detecting toxic gases in pharmaceutical, in petrochemical, and in other manufacturing units. This sensor plays an important role where there is a possibility of increased carbon dioxide levels in well-insulated buildings, which causes drowsiness and headaches to the occupants.

Electrical current-monitoring sensor: It helps to measure the real-time energy consumption. This helps in saving energy by switching off the devices when not in use and in providing an alert when there is excess surge current.

Ultrasonic sensor: Sonar (Sound Navigation and Radar) is used to estimate the distance of an object using the time taken by a sound wave to travel to the destination and back. An ultrasonic sensor is a microphone or a speaker that releases or receives ultrasound or does both. This type of sensor is useful in vehicle parking.

3.4.3 ACTUATORS

It is a machine component responsible for moving or controlling a system. To perform its operations, it requires a control signal and an energy source. In a smart building, it finds applications in locks, alarms, security cameras, solar panels, and other moving mechanisms.

3.5 IoT APPLICATIONS IN SMART BUILDINGS

Smart rest room: Smart rest room provides the following facilities:

a. Informing the housekeeping people when the level of soaps and paper towel dispenser is running low or after certain count of people use the rest room.
b. When the rest rooms start stinking badly.
c. Providing an intelligent occupancy tracker by informing when the rest rooms are occupied.
d. Providing an alert when there is a service requirement like tap leakage, no proper lighting, water shortage, handgun problems, and others.

Smart water management: The amount of water consumed on daily basis for different purposes on a global scale has increased. This has resulted in water scarcity. Hence, a check is required when water consumption levels in building increase much beyond the expectation. This can be done by installing sensors in all water supply channels in the building. The sensors will give an alert when there is excess utilization in a particular channel.

Smart building maintenance: Sensors and cameras are installed at various places around the building. The picture captured from every corner is fed to an AI tool, and if cleaning is required at a particular place, the building managers are provided with notification.

Smart parking assistance: When a new place is visited, usually a significant amount of time is wasted in finding a vacant parking space. This is usually a nuisance in crowed places. In such cases, AI plays a significant role. The pressure sensors in the ground and cameras installed all around the parking lot gather information from parking lots and feeds to AI tool. The tool then analyzes the data and provides information on the empty slots available. This information can be displayed at the entrance or make it available on the driver's phone. Navigation to the parking lot can also be provided.

Wayfinding: Wayfinding technology enables users in finding their way from one point to another in large office or corporate campus. It provides navigational assistance by guiding users to locate meeting halls, rest rooms, access elevators, or stairs. It alerts the users alternate routes when there are security breaches and fire, or in case elevators break down. It directs the fastest and safest route in case of emergency.

Asset tracking: IoT sensors can be attached to tools, vehicles, equipment, and anything that needs to be tracked. Asset tracking is being done in almost all types of industries. Before the advent of IoT, asset tracking was done manually, which was non-efficient nor scalable. It requires more man power.

The introduction of IoT leads to process automation and usage of AI in many areas of workflow that has replaced the manual work.

IoT has provided features like alerts in real time, prognostic maintenance, and real-time status of the assets by giving top-down visibility [3].

Examples:
1. Healthcare – All hospital inventory items should be tagged with RFID. The items can be scanned during entry and exit by readers placed at the entrance of the rooms. The location data is stored in a database so that administrators have accurate and latest information of all inventory in the facility.
2. Transportation and retail – In transportation and retail, it is not always about location tracking but also to have information about the condition of the shipment. In shipments, which carry perishable items like vegetables or food, there will be a requirement for adding additional monitors to check humidity and temperature along with location tracking.

Space management: IoT devices can be fit into the common spaces like meeting rooms to track the room occupancy and space utilization.

Security: IoT can be integrated into surveillance cameras, smart locks, and other security devices. These devices alert when there is any breach [4].

Fire and life safety: Latest IoT sensors with multi-sensing capabilities and miniature form can be used to identify active sources of heat that are invisible to the naked eye for monitoring electrical systems on a continuous basis. In addition, using low-power wireless solutions on independent batteries allows connections between sensors to run for many years.

It also eliminates the need for complex and dangerous wiring. IoT temperature sensors dwarf the traditional smoke detection systems by sensing fires much quicker, i.e., before they start to emit smoke. IoT data is powerful in helping combat a fire disaster or even prevent it from happening altogether.

Vibration sensing: Vibration is a critical parameter in predictive maintenance process with a central role taken by monitoring on a continuous basis. The continuous vibration monitoring of the building unearths significant information about manifold issues that lead to mechanical stress, reduction of quality, or even damage to the building in real time.

The buildings fitted with sensors and processing algorithms become intelligent, monitor their own health in real time, and become robust to extreme events.

Lighting control: Lighting system in the room is controlled based on the movement in the room using the motion detection sensor. If no movement is sensed, the lights are turned off.

Air flow control: Thermostats are installed in the room to sense the increase in room temperature.

3.6   POTENTIAL BENEFITS OF SMART BUILDING

Following are the benefits of incorporating intelligence in buildings:

1. Lighting of the building, air conditioning, and room heating are major causes for energy consumption. About 40% of energy generated in any country is consumed by buildings. By making a building smart, energy consumption can be reduced, and this has become the major concern for all the stakeholders.

2. Smart buildings provide a healthy environment by contributing to the reduction of greenhouse gas emissions.
3. Smart building facilitates the efficient use of available building space that enables companies to get the best out of the available accommodations.
4. Usage of data and analytics has enabled smart building to support interactions and knowledge sharing that intern improves business performance.
5. In smart building, the real-time data about the building operation and condition is conveyed. This helps in making decisions.
6. Smart building provides enhanced working conditions, protection, and security to the occupants.
7. The services provided by smart buildings offer a value-added functionality and help in easier operation. Both the owners and the residents relish the benefits of comfortable living and monetary gain.

3.7 DRAWBACKS OF INCORPORATING TECHNOLOGY IN BUILDINGS

As the technologies are imbibed in making smart building, it also introduces few risks as stated below:

1. IT cybersecurity risks are introduced – The interconnected systems are wide open to possible attacks. Sources for these attacks can be cyber criminals, network hackers, poor installation process, or even company employees. This leads to exposure of BMS controller on public internet, which is prone to attacks. For example, if the temperature control and defrosting action of freezers and chillers maintained in hospitals are accessible online, then any hacker can misuse it. Similarly, if CCTV footage is accessible over public internet, it might lead to evidence destruction.
2. Establishment cost is higher.
3. Data acquisition might become a challenge.

REFERENCES

[1] H. Chen, P. Chou, S. Duri, H. Lei, J. Reason, "The design and implementation of a smart building control system", in *IEEE International Conference on e-Business Engineering*, 2009.
[2] C. K. Metallidou, K. E. Psannis, E. A. Egyptiadou, "Energy Efficiency in Smart Buildings: IoT Approaches", *IEEE Access* 8, 63679–63699, 2020.
[3] M. Bajer, "IoT for smart buildings - long awaited revolution or lean evolution", in *IEEE 6th International Conference on Future Internet of Things and Cloud*, 2018.
[4] A.-E. M. Taha, A. Elabd, "IoT for Certified Sustainability in Smart Buildings," in *IEEE Network*, 2020, pp. 1–7.

4 Semi-Autonomous Human Detection, Tracking, and Following Robot in a Smart Building

Parth Mannan, Keerthi Priya Pullela,
V Berlin Hency, and O.V. Gnana Swathika
Vellore Institute of Technology

CONTENTS

4.1 INTRODUCTION

The modern world today has unfortunately become synonymous with the dangers of terrorism, militants, and the continuous threat of the military prowess of neighboring countries. Border crossings among civilians, illegal import, and export of goods etc. have also been on the rise, and despite adequate monitoring, the large areas in need of surveillance and the nature of corruption, among other factors, make it difficult for humans to effectively monitor the building in its entirety. Certain climatic conditions also prevent human surveillance in certain areas.

DOI: 10.1201/9781003240853-4

Mobile robots offer a promising solution to the problem of surveillance. Increasing popularity among mobile robots has resulted in a push for research-oriented and application-based works. A large variety of mobile robots have been used for surveillance purposes including, but not limited to, the likes of unmanned aerial vehicles, all-terrain vehicles, bio-inspired robots, and underwater vehicles.

The dynamic nature of human movement renders a static surveillance system less useful as the object of interest may leave the field of view before appropriate authorities can take action. The system therefore needs to not only detect any human being in the frame but also have the ability to track and follow its movement to a certain extent.

This paper describes the design, implementation, and experimental evaluation of such a mobile surveillance system. The system has been tested on a robust ground vehicle. The system performs image processing on an embedded computer to navigate, detect, and track human beings. The mobile robot is controlled by a secondary embedded computer to ensure the real-time tracking and movement of the robot. The system has also been included with a password-protected lock, which can be used to reset the robot to its initial state once the authorities reach the detected human.

The paper is organized as follows: Section 4.2 is a discussion on similar works and other literature survey. Section 4.3 describes the methods used to develop the surveillance system. Section 4.4 discusses the results obtained from the work done. The paper is concluded in Section 4.5 by discussing the scope for future work.

4.2 RELATED WORK

4.2.1 HUMAN DETECTION

Awad and Shamroukh (2014) have used a feed-forward neural network-based method for human detection. A passive infrared sensor (PIR) is initially used in the system to detect the existence of human body parts and then triggers the camera feed on successful detection. Relatively small number of images can be acquired in this method and processed during the operation; this considerably reduces the cost of image processing, power consumption, and data transmission. However, this system is not equipped with a 360° field of view, reducing the effectiveness of the surveillance systems. Second, the system does not intend to track the human after detection, which is integral to having effective surveillance applications especially in areas like smart building.

Naveen et al. (2014) use a PIR sensor-based detection system in their work. The system uses GSM to communicate with security personnel and triggers the camera on an Android-based smartphone to capture an image and send that to appropriate authorities by e-mail. However, the presence of a reliable GSM architecture is usually unavailable on all buildings. Additionally, the sole use of a PIR sensor for human detection is unreliable as any kind of motion; human or animals etc. will be detected by the system. The works of Sravani et al. (2014), Trupti et al. (2014), Purnima et al. (2014), and Brem et al. (2015) also suffer from similar limitations of relying solely on the PIR sensor for the purpose of human detection. Sandeep et al. (2011) use an ultrasonic sensor to detect motion along with temperature sensor to detect fire and metal sensor to detect the possibility of bombs in the building. Nevertheless, the use of only an ultrasonic sensor to determine human presence is not a very effective and reliable solution.

Correa et al. (2012) use thermal as well as visual sources of information for human detection. The presence of humans is detected using the face detector, and to identify them, a face recognition system is used. If no direct identification is possible, then to improve the relative pose of a candidate object/person, an active vision search mechanism is employed.

4.2.2 Object Tracking

Babenko et al. (2011) aimed to build a tracker based on multiple instance learning (MIL) appearance model and training it in an online manner. The work goes on to propose an efficient algorithm called MILTrack, and its performance is compared with two other algorithms: first is an older algorithm based on AdaBoost, trained in an online manner, and second is an algorithm called FragTrack, which uses static appearance models based on integral histograms. The results show that MILTrack outperforms the others and it was later included in the Tracking API of the OpenCV library.

Joao et al. (2014) proposed an analytic model for datasets of thousands of translated patches. For kernel regression, a new kernelized correlation filter (KCF) that has the same complexity as its linear counterpart unlike other algorithms is proposed. Moreover, a fast multi-channel extension of linear correlation filters, via a linear kernel, which they call "dual correlation filter" (DCF), is also proposed. Both KCF and DCF outperform compared to the old trackers like Struck and TLD. Their tracking framework is made open-source and included in the OpenCV Tracking API. However, the tracking algorithm gives slower results as compared to MIL when used with the proposed system.

Kalal et al. (2010) proposed a novel tracking framework, known as TLD, that decomposes the long-term tracking task into tracking, learning, and detection. The tracker is made to follow the object from frame to frame. The detector localizes all appearances that are being observed and corrects the tracker if it is necessary. The learning estimates the possible detector's errors and updates it to avoid these errors in the future. The learning process is modeled and named as a discrete dynamical system. There are a few limitations for this algorithm. For instance, in case of full out-of-plane rotation, TLD does not perform well and it is made to train only the detector and the tracker stays fixed. As a result, the tracker repeats the same errors.

Optical flow is the pattern of apparent motion of image objects between two consecutive frames, which is caused by the movement of camera or object. It is 2D vector field where each vector is a displacement vector showing the movement of points from the first frame to the second. Lucas-Kanade (LK) method takes a 3×3 patch around a single pixel in the region of interest, so all the nine points have the similar motion trajectory and form optical flow equations. These equations are solved by the least-squares fit method. In OpenCV, on the first frame, Shi-Tomasi algorithm is used to detect corner points. These points are iteratively tracked LK optical flow. LK algorithm has been explained in detail in the further sections. Implementation of this algorithm in OpenCV is based on the paper proposed by Bouguet (2010).

4.2.3 SURVEILLANCE ROBOTS

In the field of Urban Search and Rescue (USAR), a significant amount of work has been done in the direction towards making robots autonomous, robust, and capable of traversing difficult and unpredictable terrains. Though the application of USAR robots is different from human detection within the war field, it still provides a significant insight on autonomous surveillance/search and rescue robots.

Remote Operated and Controlled Hexapod (ROACH) by Capello et al. (2005) is a six-legged design that has significant advantages over wheeled designs. It has a camera to transmit live images and video to the receiver; however, it does not aim to detect humans autonomously.

Kohga is a snake-like robot developed by Amerada et al. (2004); the mechanical structure of which allows it to crawl through narrow spaces, which offer a significant advantage in the area of USAR.

Burion (2004) presents a work that provides a sensor suite to detect human for the USAR robots. This study compared and evaluated several types of sensors for human detection, such as pyroelectric sensor, USB camera, microphone, and IR camera. The study also involved a human detection robot but it is not fully autonomous and is dependent on human administrator.

Md Jahidul Islam et al. studied about an autonomous robot to follow its human companion in applications such as search and rescue operations, industry, healthcare, and social interactions.

4.3 PROPOSED WORK

In this section, the system architecture of the surveillance system and its working algorithm has been discussed. The system uses two embedded computers that communicate with each other using an inbuilt UART along with a USB webcam to capture image data for processing.

4.3.1 HARDWARE DESIGN

The block diagram of the proposed system is shown in Figure 4.1. It houses a Raspberry Pi 3 board as its primary processor and a TI Launchpad MSP-EXP430G2553 board as a secondary processor, the reasons for which are explained in Section 4.3.2. A peripheral webcam attached to the Raspberry Pi processor is responsible for capturing image data for surveillance. The USB webcam is fixed on a stepper motor to allow the robot to see 360° around itself to perform surveillance. A matrix keypad has also been attached as a user interface for password entry.

4.3.2 SOFTWARE DESIGN

The software architecture of the system is built on a Linux core Raspbian Jessie. The major tool used for image processing is OpenCV 3.2.0 and has been programmed using Python 3.4.2. Image data is constantly collected and processed by the system. Upon the successful detection of human presence in an image, the system enables

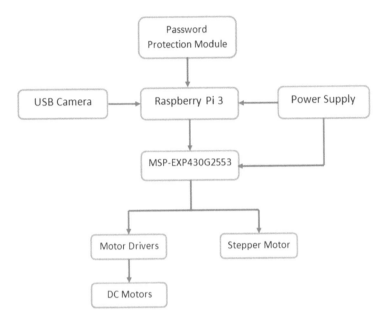

FIGURE 4.1 Block diagram of the system.

the tracking and following module. The Raspberry Pi also sends this information to the Launchpad to align the direction of the camera and robot so as to place the detected human in the center of the frame. This is done as the direction of movement of the detected human is unpredictable. The tracking data is constantly sent to the Launchpad, which controls the robot and camera movement in order to follow the movement of the detected human. The Raspberry Pi also checks for password entry simultaneously and resets the robot on completed authentication. The algorithm design for detection, tracking, and following has been explained in the next subsection in detail.

4.3.3 WORKING ALGORITHM

In this subsection, the algorithms used for human detection, tracking, and following have been described in detail. The algorithms are shown in Figures 4.2 and 4.3.

1. The first challenge is to successfully detect the presence of human being in the frame. The system uses histogram of oriented gradients (HOG) along with a support vector machine (SVM) classifier to detect the presence of humans.
2. Once detected, a region of interest is built around the detection. After communicating this region of interest to the secondary processor, the primary processor enters the tracking mode.
3. The tracker has been built based on the MIL and LK optical flow trackers. As elaborated in Section 4.2, the MIL and LK are not self-sufficient. The optical flow points are attributed inside the region of interest tracked and defined by MIL tracker.

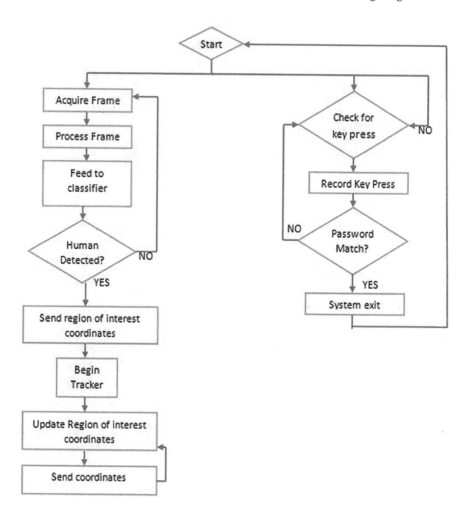

FIGURE 4.2   Working algorithm of the primary processor.

4. It is a safe assumption to make that a large majority of the optical flow points define the edge of the human body. The average of the co-ordinates of all the points is communicated continuously to the secondary processor, which decides – based on these averages – the motion of the detected human.
5. The averages in the X-direction and the Y-direction are stored separately in a running average window of ten samples implemented using a First in First out (FIFO) queuing mechanism.
6. The data points in the running window are averaged to form the basis of the motion decision by the secondary processor. This is done to filter out bad

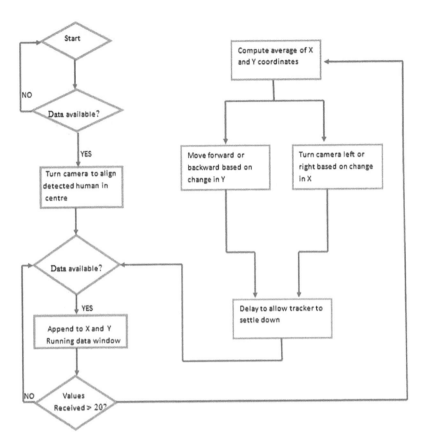

FIGURE 4.3   Working algorithm of the secondary processor.

data arising due to errors in communication, absence of a minimum number of points in the LK tracker due to poor edge detection, and lighting conditions among other factors.

7. The averages in the X-direction define the movement as left, right, or central, and the averages in the y-direction define the movement as forward, backward, or stationary.

4.4   RESULTS AND DISCUSSION

4.4.1   EXPERIMENTAL SETTING

The system was tested on an on-ground vehicle in an open environment. The structure of the vehicle used for testing is shown in Figure 4.4. The structure built houses the primary and the secondary processor with the power supply inside the body in

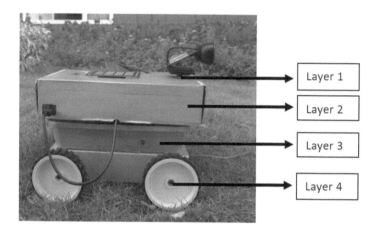

FIGURE 4.4   Mechanical structure of the ground vehicle.

layers 2 and 3 of the structure. The camera has been attached to a stepper motor to control the movement of the camera. The camera body and the keypad have been mounted outside the body of the robot on the layer 1 of the structure. Layer 4 comprises the motors and the wheels that are responsible for the movement of the robot.

4.4.2   HUMAN DETECTION RANGE TEST

The human detection module of the system based on HOG was put through a range test by testing the module on varying distances of human from the camera. Selective test results are shown in Figure 4.5. It was observed that in open environments under average lighting conditions, human detection module worked satisfactorily up to a distance of around 60 ft and beyond that the behavior is unpredictable.

4.5   HUMAN TRACKING SPEED TEST

The human tracking module of the system using MIL tracker along with LK optical flow tracker was tested for accuracy against the speed of the human, the system is tracking whom the system is tracking. The processing time on each obtained frame induces a lag, which may result in the human getting lost from the frame. Figure 4.6 shows the case where the tracker failed to follow the human.

4.6   HUMAN TRACKING OCCLUSION RECOVERY TEST

The human tracking module was also tested against occlusions. The object being tracked was intentionally blocked in the frame to simulate hindered environments that could be present in the war field. The ability of the system to recover from losing the object to its reappearance was tested. The occlusion test is shown in Figure 4.7.

FIGURE 4.5 Human detection range test.

4.7 COMPARISON OF TRACKING ALGORITHMS

A few different tracking algorithms were compared for performance against a sample video against speed of processing and accuracy. The results from the test are shown in Figure 4.8.

FIGURE 4.6   Tracker loses target.

FIGURE 4.7   Tracker recovering from occlusions.

Table 4.1 summarizes the comparisons made between different algorithms. The three parameters used to compare are:

1. Processing delay
2. Tracking accuracy
3. Occlusion recovery.

4.8   CONCLUSION AND CHALLENGES AHEAD

This report describes the design, associated algorithms, implementation, and experimental evaluation of a human detection, tracking, and following system. The system uses HOG gradients with SVM trainer to detect a human body in the frame. MIL-based tracker along with LK optical flow tracker has been used in the tracking module. A comparison of different tracking modules was made and presented. The real-time experimental evaluation of the system was carried out in daylight in open environment. Evaluations show that the successful rate of human detection was 92%. The detection module was also tested with varying distances to the human and performed satisfactorily up to a distance of 60 ft. The tracking module designed and

FIGURE 4.8   Comparison of tracking algorithms.

TABLE 4.1
Comparison of Tracking Algorithms

Tracker Algorithm	Processing Delay	Tracking Accuracy	Occlusion Recovery
KCF	Medium	Medium	Partial
TLD	Low	Low	High
MIL	High	High	Partial
MedianFlow	Very high	Low	None
Boosting	Medium	Low	Low
Camshift	Medium	Very low	High
LK	Very low	Medium	High
Proposed MIL + LK	High	Very high	Partial

implemented was also observed to perform satisfactorily in tracking a human walking with a maximum speed of 50 m/minutes approximately. Furthermore, the system built has an accumulated cost of only 210 USD, making it an extremely suitable system for deployment on a large scale in the war field or other surveillance applications.

The speed at which the tracking module can effectively function can be significantly improved by using a more powerful on-board system than the Raspberry Pi 3. The addition of a GPU can also lead to significant improvements in speed. The system can be supplemented by using two cameras instead of one to provide stereo vision. Stereo vision can provide the ability to measure depth to a detected object making following more accurate. The system can also be made autonomous if supplemented with a GPS to help the system identify its starting position and the surveillance path.

REFERENCES

Amerada, T., Yams, T., Igarashit, H., Matsunos, F., 2004. "Development of the snake-like rescue robot KOHGA", IEEE, pp. 5081–5086.

Awad, F., Shamroukh, R., 2014. Human detection by robotic urban search and rescue using image processing and neural networks. *International Journal of Intelligence Science*, 4, 39–53.

Babenko, B., Yang, M., Belongie, S., 2011. Robust object tracking with online multiple instance learning, *IEEE Transactions on Pattern Analysis and Machine Intelligence*.

Bhatia, S., Dhillon, H.S., Kumar, N., 2011. Alive human body detection system using an autonomous mobile rescue robot, *2011 Annual IEEE India Conference*, Hyderabad, pp. 1–5.

Bouguet, J.-Y., *Pyramidal Implementation of the Affine Lucas Kanade Feature Tracker Description of the Algorithm*, Microprocessor Research Labs, Intel Corporation.

Cappello, C., Olsen, C., Auen, M., 2005. *Remote Operated and Controlled Hexapod (ROACH) Robot*, Rescue Robot League Competition Atlanta, Georgia.

Correa, M., et al., 2012. Human detection and identification by robots using thermal and visual information in domestic environments, *Journal of Intelligent & Robotic Systems*, 66, 223–243.

http://docs.opencv.org/trunk/d7/d8b/tutorial_py_lucas_kanade.html.

Islam , M.J., Hong, J., Sattar, J., 2018. "Person following by autonomous robots: a categorical overview", *The International Journal of Robotics Research*, 38, 12, 1–32.

Joao, F., et al., 2014. High-speed tracking with kernelized correlation filters, *IEEE Transactions on Pattern Analysis and Machine Intelligence*, 37, 3, 583–596.

Kalal, Z., et al., 2010. "Tracking-learning-detection", *IEEE Transactions on Pattern Analysis and Machine Intelligence*, 6, 1.

Kumar, M.B., Manikandan D., Gowdem, M., Balasubramanian, D., 2015. Mobile phone controlled alive human detector using robotics. *International Journal of Advanced Research in Computer Engineering & Technology*, 4(3).

Naveen, C.M., Ramesh, B., Shivakumar, G., Manjunath, J.R., 2014. Android based autonomous intelligent robot for border security. *International Journal of Innovative Science, Engineering & Technology*, 1(5).

Purnima, G., Aravind, S., Varghese, R.M., Mathew, N.A., Gayathri, C.S., 2014. Alive human body detection and tracking system using an autonomous Pc controlled rescue robot. *International Journal of Emerging Technology and Advanced Engineering*, 4(12).

Sravani, K., Ahmed, M.P., Sekhar, N.C., Sirisha, G., Prasad, V., 2014. Human motion detection using passive infra red sensor. *International Journal of Research in Computer Applications & Information Technology*, 2(2), 28–32.

Steve B. 2004. "Human detection for robotic urban search and rescue", Info science database of the publications and research reports. Technical Report.

Trupti, B., Satyanarayan, R., Mukhedkar, M., 2014. Mobile rescue robot for human body detection in rescue operation of disaster. *International Journal of Advanced Research in Electrical, Electronics and Instrumentation Engineering*, 3(6).

5 Current-Starved and Skewed Topologies of CNTFET-Based Ring Oscillators for Temperature Sensors

Gangachalam Akula and Reena Monica P
Vellore Institute of Technology

CONTENTS

5.1 INTRODUCTION

The IC technology using metal oxide semiconductor faces many obstacles as we venture into the nanometer range. The factors that influence the MOSFETs are associated leakage currents, static power dissipation, and short channel effects. To overcome these defects, carbon nanotube field-effect transistors (CNTFETs) having single-walled carbon nanotubes (SWCNTs) are proposed as future-generation devices since their discovery in 1991. Carbon tubes are hexagon structures that are made up of benzene. The virtual source CNTFET (VS CNTFET) consists of only three terminals, and there is no body terminal.

Performance and characteristics of CNT technology are more efficient compared to those of FINFET and conventional CMOS technology. CNT provides a unique mechanism to control the electrical and mechanical propertied of threshold voltage by varying the chirality vector (m, n) [1]. So, we need to consider the parameters carefully to come up with an efficient design and improved performance in CNTFET technology devices. The threshold voltage of CNTFET depends upon chirality and diameter of the devices. The diameter (D_{CNT}) and the threshold voltage (V_{th}) of CNTFETs are calculated based on equations (5.1) and (5.2):

$$V_{th} \approx \frac{E_g}{2e} = \frac{aV_\pi}{\sqrt{3}eD_{CNT}} \tag{5.1}$$

where $\alpha = 2.49\,\text{Å}$ and $V_\pi = 3.033\,\text{eV}$ is the carbon $\pi-\pi$ bond energy.

$$D_{CNT} = \frac{\sqrt{3}a_o\sqrt{n^2 + m^2 + nm}}{\pi} \tag{5.2}$$

where $a_o = 0.142$ is the center-to-center distance between carbon atoms.

5.2 BACKGROUND

Based on oscillator classification, LC oscillator [2] can be widely used for the betterment of noise performance. Parameters like speed, tuning range, and area-wise non-resonant oscillators are good. In the non-resonant network family, ring oscillators have good characteristics when compared with other oscillators. Figure 5.1 shows the structure of ring oscillator design. In the design, an odd number of inverters are connected to the feedback network in a loop, because the input value must be inverted at the output, i.e., low to high or high to low sinusoidal waveform. The frequency of

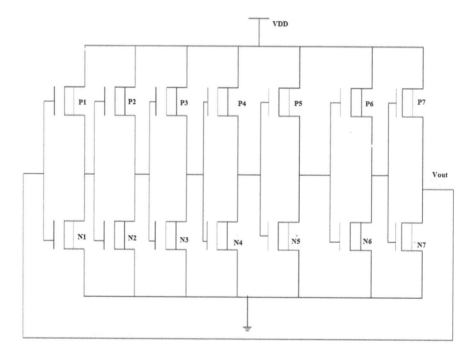

FIGURE 5.1   Seven-stage CNTFET ring oscillator.

each oscillator depends upon the number of serial stages (N), and the propagation delay of each stage of inverter (t_{delay}) is given by equation (5.3):

$$f_{osc} = 1/2Nt_{delay} \tag{5.3}$$

In the ring oscillators, there are different characteristics like current-starved, skewed topology, single- and dual-stage delays, subthreshold design, and quadrature and digitally controllable switches. Figure 5.2 shows the waveform of ring oscillators. Changing the output frequency in a ring oscillator is not possible because the input is connected in a feedback form [3]. Hence to adjust the frequency of oscillations, the current-starving technique is developed. In this method, to adjust the oscillation frequency, an extra driver circuit, which is a voltage-controlled source, is given as input to the oscillator circuit with a 0.8 V power supply. The supply voltage provides an output voltage swing to provide acceptable phase noise performances. Skewed topologies can be classified into high-skewed and low-skewed topologies. In high-skewed topologies, we use p-type CNTFET, since it feeds earlier than the n-types. Similarly, in low-skewed topologies, it is an n-type CNTFET because it feeds earlier than the p-types.

In this work, a ring oscillator having different characteristics using VS CNTFET has been proposed. Characteristics of ring oscillator [4], such as current-starved, skewed topologies of driver, load, and current-controlled sources performance, have been observed. Based on the performance of the circuit, a new characteristic, current-mode injection-locked loop (C-IFLM) is designed, and they have trans receivers, which are mainly used in biomedical sensor applications and in smart buildings.

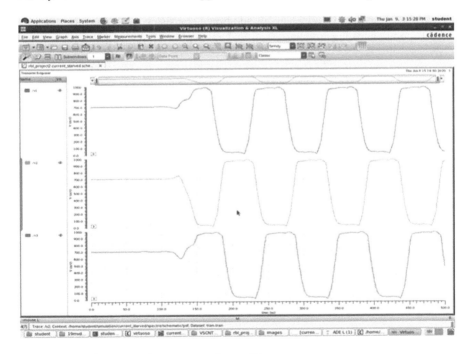

FIGURE 5.2 Waveform of seven-stage ring oscillator.

5.3 CURRENT-STARVED RING OSCILLATOR

Figure 5.3 shows the structure of current-starved (CS) technique [5]. It can be classified into three types based on driver, load, and current-controlled source. In the driver part, the circuit consists of several inverter stages. In each stage the transistor of both PMOS and NMOS will be connected above and below, each having one more PMOS and NMOS transistor like in dynamic circuit topology. But in the techniques using CNTFET instead of a clock signal having pre-charge and evaluation phase, we should use a driver circuit, which acts as a controlling source. The voltage-controlled source varies the tuning frequency. In the load part, [6] due to the serial chains connection, it will consumes less power than the driver part of the circuit but it will not have any controlled source; hence, the frequency cannot be changed. In the current-controlled source, we have to change the values of the current because the current will be directly proportional to the chirality vectors. For low power consumption, the biasing current should be as low as possible. Therefore, the current-controlled source will consume less power than the other two.

The leakage power consumed by the voltage-controlled oscillator (VCO) circuit is given by equation (5.4):

$$P_{leakage} = V_{DD} \times \sum I_{leakage} \tag{5.4}$$

where $I_{leakage}$ is the penetrating leakage current in the turn-OFF transistors.

V_{DD} is the supply voltage.

Figure 5.4 shows the waveform of the current-controlled and the voltage-controlled source for an inverter stage where an additional transistor is required for dynamic circuit operation having pre-charge and evaluation phase. But here the circuit will not

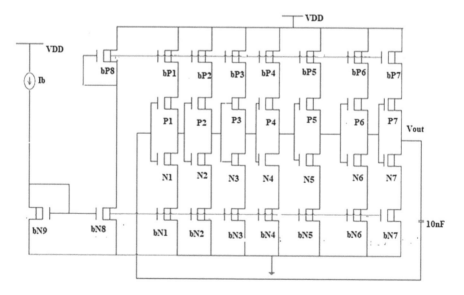

FIGURE 5.3   Current-starved, current-controlled oscillator.

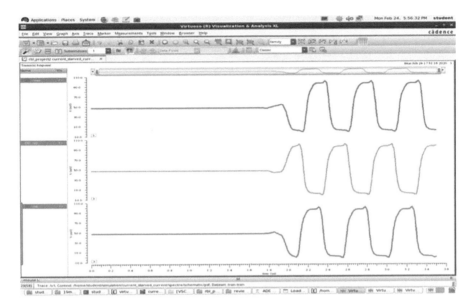

FIGURE 5.4 Waveform of current-starved, current-controlled oscillator.

have clock phase cycles; instead, we will use a differential circuit-like amplifier that acts as the driver. Sinusoidal representation will be there because of phase difference between the signals.

5.4 SKEWED TOPOLOGIES

In skewed topologies, there are two types of classification: high skew and low skew, for which the performance of the three parameters can be observed. The operation of three types of skewed topologies is the same as that of current-starved technologies [7], like the driver, load, and current-controlled source. In these skewed topologies classification, the high-skew operation is performed like in the ring oscillator connection, where the feedback is connected to PMOS devices (i.e., p-type devices connected before the n-type devices) to increase the transition rate at the output. Similarly, for low-skew topologies, the feedback path is connected to NMOS devices (i.e., n-type devices are fed earlier than p-type devices) to increase the transition rate at the output.

5.5 NEGATIVE PMOS AND NMOS SKEW RING OSCILLATOR

In these topologies of the circuit, the PMOS transistors are connected to the negative feedback elements, and whenever a low to high transition occurs, the performance of NMOS is found [7] to be lower than that of PMOS. Figures 5.5 and 5.6 show the structure and waveform of PMOS skew ring oscillator. Similarly, when a high to low transition occurs, PMOS will be lower than NMOS. Similarly, when connected by a negative delay element, the operation is the same except that the transistors will be replaced. Figures 5.7 and 5.8 show the structure and waveform of NMOS skew ring oscillator.

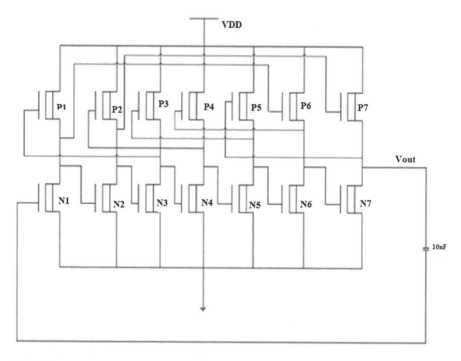

FIGURE 5.5   Negative PMOS-skewed oscillators.

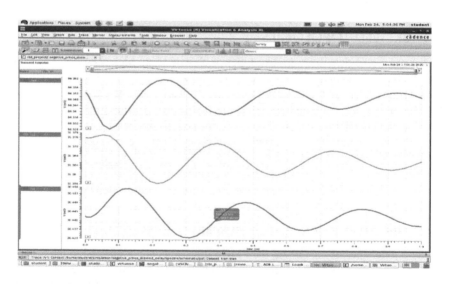

FIGURE 5.6   Waveform of negative PMOS-skewed oscillator.

The NMOS will be faster than PMOS, because the mobility of electrons will be more compared to the mobility of holes ($\mu_n > \mu_p$), and hence, power consumption is also more in both these skewed ring oscillators.

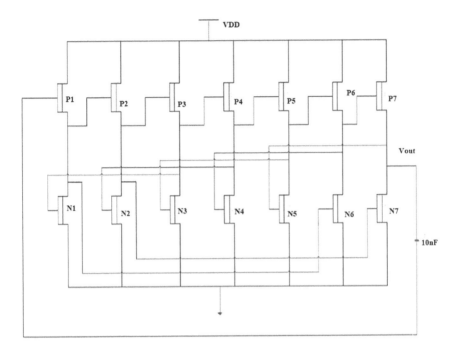

FIGURE 5.7   Negative NMOS-skewed oscillator.

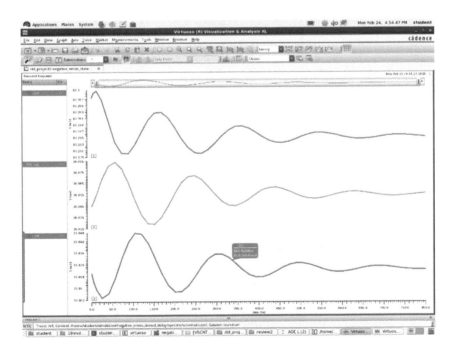

FIGURE 5.8   Waveform of negative NMOS-skewed oscillator.

5.6 CURRENT-CONTROLLED NMOS AND PMOS RING OSCILLATORS

Figures 5.9 and 5.10 show the structure of current-controlled NMOS and PMOS ring oscillators. In the current-controlled ring oscillators, the PMOS and NMOS circuit operation is the same as that of the negative NMOS and PMOS skew operation but there will be an additional current-controlled source [8], which will act as the driver part. These approaches show an improvement in the frequency of oscillation, but the tradeoff is in power consumption. Combining driver part and negative circuit will be helpful in low power consumption and high frequency of operation of negative skewed delay approach. The operation is similar to that of current-starved current-controlled source. Figures 5.11 and 5.12 show the simulation waveform of current-controlled NMOS and PMOS ring oscillators.

Based on these observations, the power consumption and high-frequency operation, current-controlled source is comparable. Therefore, based on the observed high-performance metrics, an application related to low power current starved using harmonic injection CNTFET-based ring oscillator was developed.

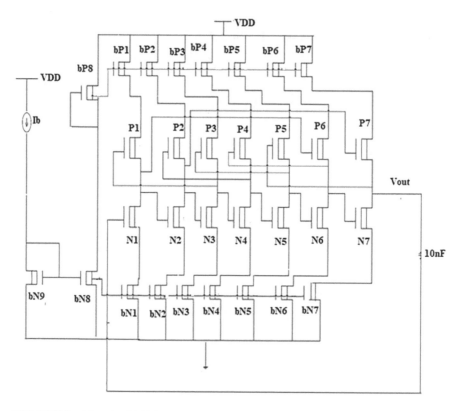

FIGURE 5.9 Negative PMOS current-controlled source.

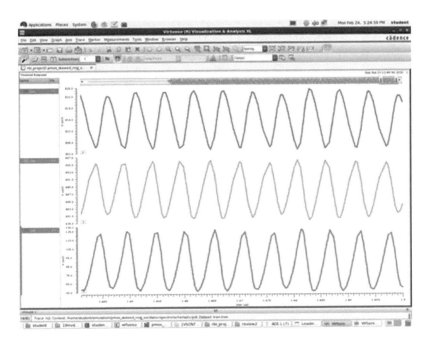

FIGURE 5.10 Negative NMOS current-controlled source.

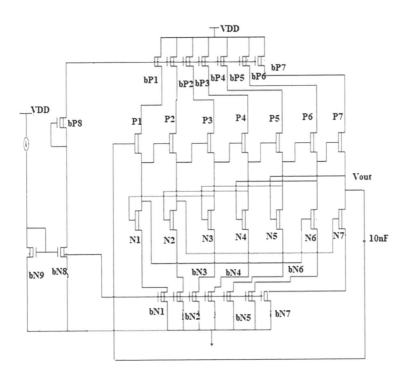

FIGURE 5.11 Waveform negative PMOS current-controlled source.

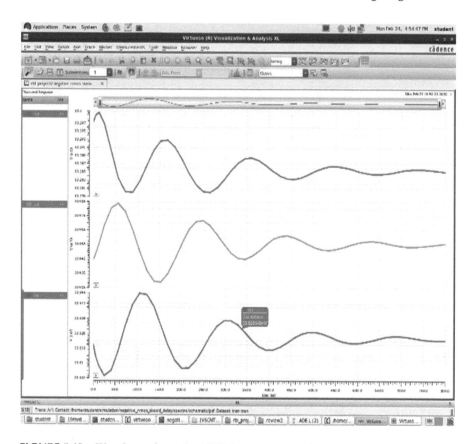

FIGURE 5.12   Waveform of negative NMOS current-controlled source.

5.7 CURRENT-MODE CNTFET IFLM (C-IFLM)

In this model, the C-IFLM is proposed with the frequency multiplication method by using frequency divider where the reference signal is multiplied [9] to the number of times the frequency of multiplication. Because of the frequency of signal which is multiplied to the subharmonics injection frequency of multiplier, the free-running frequency of reference signal is insensitive to current, so the current variation affects the temperature.

Figures 5.13 and 5.14 show the structure and waveform of C-IFLM. In the proposed method, a reference signal of frequency 500 MHZ is applied to the duty cycle controller and bias circuit followed by three stages of inverter system. Low-cycle current pulse duty cycle reduction method and then cell liberation are also applied with a reference frequency. This application of C-IFLM is used in temperature sensors. The oscillator is locked to increase harmonic oscillation for better performance (Tables 5.1 and 5.2).

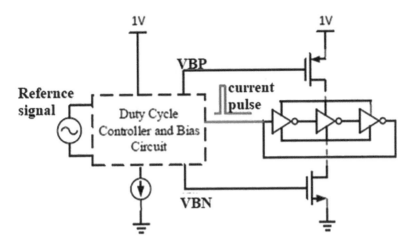

FIGURE 5.13 Current-mode CNTFET IFLM.

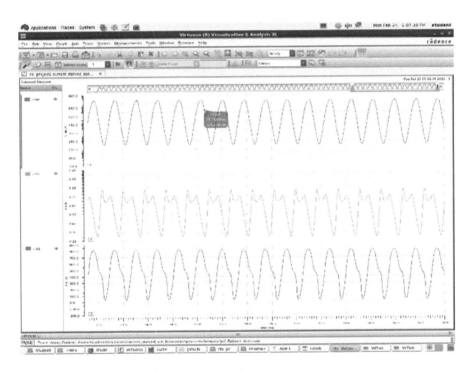

FIGURE 5.14 Waveform of C-IFLM harmonic oscillator.

TABLE 5.1
Power and Energy Values of Load

Types of Ring Oscillator	Voltages	Voltage Values (mV)	Average Power (W)	Energy (µJ)
Current-starved load	V3	84.575	82.3e-3	4.09
	V2	45.903	44.7e-3	2.25
	V1	133.245	98.35e-3	4.89
Negative NMOS-skewed delay	V3	83.29	83.29e-3	3.31
	V2	30.98	30.96e-3	1.23
	V1	33.84	33.82e-3	1.34
Negative PMOS-skewed delay	V3	84.36	84.35e-3	3.35
	V2	31.37	31.37e-3	1.24
	V1	30.45	30.44e-3	1.21

TABLE 5.2
Power and Energy Values of Current-Controlled Source

Types of Ring Oscillator	Voltages	Voltage Values (mV)	Average Power (W)	Energy (µJ)
Current-starved load	V3	100.96	43.026e-3	8.12
	V2	100.96	48.94e-3	8.52
	V1	100.45	46.13e-3	8.03
Negative NMOS-skewed delay	V3	92.33	60.74e-3	7.56
	V2	89.59	41.50e-3	5.16
	V1	89.17	42.32e-3	5.26
Negative PMOS-skewed delay	V3	82.77	82.9e-3	6.19
	V2	45.34	48.7e-3	3.5
	V1	123.64	89.94e-3	6.71

TABLE 5.3
Power and Energy Values of Driver

Types of Ring Oscillator	Voltages	Voltage Values (mV)	Average Power (W)	Energy (µJ)
Current-starved load	V3	946.44	536.9e-3	2.67
	V2	975.13	526.0e-3	2.61
	V1	975.13	544.0e-3	2.70
High-skew topologies	V3	936.62	506.0e-3	2.51
	V2	953.13	527.6e-3	2.62
	V1	945.11	530.6e-3	2.67
Low-skew topologies	V3	930.8	529.1e-3	2.63
	V2	951.8	545.1e-3	2.71
	V1	958.12	539.1e-3	2.68

TABLE 5.4
Power and Energy Values of Subharmonics

Voltage	Voltage Values (mV)	Average Power (W)	Energy (µJ)
V3	738.555	738.55e-3	1.10
V2	810.594	810.59e-3	1.21
V1	821.472	486.9e-3	0.72

5.8 SUMMARY

In this work, comparison of three categories of ring oscillator characteristics for current-starved and high-skew and low-skew parameters is performed. Based on the simulation results, we conclude that the skewed topologies are suitable for high-frequency application, whereas the current-starved characteristics are suitable for voltage swing application, and the current-controlled source is best for low-power circuits. Deriving from the design of current-starved, high skew, and low skew, we designed a CNTFET IFLM based on the injection-locked subharmonics oscillator by using the frequency multiplication of the circuit. Tables 5.1–5.4 show the power and energy values of load, current-controlled source, driver, and subharmonics, respectively, calculated using Cadence Virtuoso. The average power for current-starved, high skew, and low skew for driver, load, and controlled source is 26.73, 4.09, and 2.43 mW, respectively. The transceivers are used in temperature sensors that are employed in smart buildings. The proposed ring oscillator can be used for temperature sensors. The CMOS-based temperature sensors have less resolution as they operate with bandgap reference. The proposed sensors will occupy relatively less silicon area.

REFERENCES

[1] C. Li, J. Lin, A 19 Ghz linear-wide-tuning-range quadrature ring oscillator in 130nm cmos for non-contact vital sign radar application, *IEEE Microw. Wirel. Compon. Lett.* 20 (1) (2010) 34–36.

[2] G. Li, L. Liu, Y. Tang, E. Afshari, A low-phase-noise wide-tuning-range oscillator based on resonant mode switching, *IEEE J. Solid State Circuits* 47 (6) (2012) 1295–1308.

[3] H. Borjkhani, S. Sheikhaei, M. Borjkhani, Low power current starved sub-harmonic injection locked ring oscillator, in *2014 22nd Iranian Conference on Electrical Engineering (ICEE)*, IEEE, 2014, pp. 38–42.

[4] K. Hassanli, S.M. Sayedi, R. Dehghani, A. Jalili, J.J. Wikner, A low-power wide tuning-range cmos current-controlled oscillator, *Integr. VLSI J.* 55 (2016) 57–66.

[5] S. Suman, M. Bhardwaj, B. Singh, An improved performance ring oscillator design, in *2012 Second International Conference on Advanced Computing & Communication Technologies (ACCT)*, IEEE, 2012, pp. 236–239.

[6] A.K. Mahato, Ultra low frequency cmos ring oscillator design, in *2014 Recent Advances in Engineering and Computational Sciences (RAECS)*, IEEE, 2014, pp. 1–5.

[7] A. ElMourabit, G.-N. Lu, P. Pittet, Y. Birjali, F. Lahjomri, M. Zhang, *A new method to enhance frequency operation of cmos ring oscillators*, Int. J. Electron. 99 (3) (2012) 351–360.

[8] S. Park, C. Min, S. Cho, A 95nw ring oscillator-based temperature sensor for rfid tags in 0.13 m cmos, in *2009 IEEE International Symposium on Circuits and Systems*, IEEE, 2009, pp. 1153–1156.

[9] N. Pandey, R. Pandey, T. Mittal, K. Gupta, R. Pandey, Ring and coupled ring oscillator in subthreshold region, in *2014 International Conference on Signal Propagation and Computer Technology (ICSPCT)*, IEEE, 2014, pp. 132–136.

6 Efficient Data Storage with Tier IV Data Center in Smart Buildings

*Dishant Bhagdev, N. Srinivasan,
and R. Balakrishnan*
L&T Construction

CONTENTS

6.1 INTRODUCTION

The smart building realization is progressing with the advancements on the Internet of Things (IoT). There are thousands of devices connected in the building, which interact with each other and to the servers; the base of the IoT lies in the data collection from all the devices, which process the data collected, analysis, and feedback after the analysis. The amount of data traffic generated is massive in TBs, which first needs to be transported to the distant cloud or servers in real time for storage along with ultra-high-speed computation. The solution is building-integrated data center, which can increase the efficiency and reduce the burden on the server [1]. Smart Building Data Center (SBDC) has a capacity to provide superior connectivity and low network congestion with savings in energy and capital.

Tier IV configuration of data center is the right solution for the important data, which demands almost zero power outage and requires the highest level of protection. The Tier IV data center is completely redundant and fault tolerant. All the equipment or devices should be fully redundant (N + N), which is mission critical and serving the racks. Any single point of failure in Tier IV data center cannot affect the data center operation, which means that no critical equipment operation should be affected by any fault. The standard referred for data center is given by **UPTIME INSTITUTE** [2]. The uptime institute is the internationally recognized organization, which has standards for Tier classifications and certifies the data centers accordingly. This paper explains the design parameters and criteria of Tier IV data center.

DOI: 10.1201/9781003240853-6

6.2 DESIGN CRITERIA FOR TIER IV DATA CENTER

The backbone of data center is racks or servers, where all the data is stored, and the entire design criteria revolve around the racks and the equipment critical for the operation of the racks. The design process to be followed in order to establish a new data center is:

i. Estimation of equipment telecommunications, space, power, and cooling requirement of the data center at full capacity, including future requirements.
ii. The architects and engineers shall be provided with the main requirements and other support area requirement.
iii. Coordination of the preliminary space plans of the data center.
iv. Preparation of floor-wise equipment layouts and distribution schemes, including mechanical and electrical-pertaining equipment according to full capacity.
v. Design of telecommunication cabling system based on the existing and planned future equipment located in the data center.

The Tier Gap Analysis (TGA) service by uptime institute confirms the performance requisites by Tier Standard Topology on reviewing the following items:

- Basis of design – architecture and approach
- Electrical schematic diagrams – single line diagrams
- Mechanical diagrams – flow diagrams and air flow diagrams
- Equipment layout drawings
- Equipment specifications
- Load calculations – electrical power demand and cooling demand
- Control system design and operations
- Sequence of operations, operational plans
- Architectural plans
- As well as any other documents required to ensure that all aspects of the design meet the requirements of the Tier objective.

The detailed design criteria for Tier IV Data center shown in Figure 6.1 can be understood based on the discipline-wise compliances to the Tier IV uptime requirements.

a. Architectural Design
- Location – Rooms should not have exterior windows, as exterior windows increase heat load and reduce security.
- Size – Size should include proper clearances to meet all the current requirements and the projected future requirements.
- Ceiling height – The minimum clear height shall be 2.6 m (8.5 ft) from the finished floor to any obstruction.
b. Electrical
- Power – The data center substation should be fed by two different sources of supply.

- Standby power – All the mission-critical equipment should be supported by UPS, and UPS should be supported by a generator. The generator should be adequately sized to handle the full load, including losses.
- UPS containing flooded cell batteries shall be located in a separate room, which is immediately adjacent.
- Equipment redundancy – All the equipment serving to the data center operation should be redundant.
- As shown in Figure 6.1, at each point, the system is completely redundant, which means for every equipment A (Active-1), there will always be equipment B (Active-2); both the equipment shall be equally (50%) loaded; and both should be able to take 100% load.
- Hence at any point of time, any equipment failure will never affect data center operation.
- The equipment A (Active-1) and equipment B (Active-2) shall be in two different rooms, and those rooms shall be in different fire compartments.
- The distribution shall be planned such that the equipment A (Active-1) and equipment B (Active-2) should feed the racks in two different paths.
- The changeover of the critical equipment shall be operated by means of ATS (automatic transfer switch).

c. Fire Protection System
- The fire protection system and hand-held fire extinguisher shall comply with NFPA-75 or applicable regional code for fire extinguishers used in data centers.
- Fire sprinkler system shall be pre-action type while proposing to data hall.
- The cable types and the fire suppression practices shall be considered that focus on minimizing the damage of equipment and facility in the event of fire.
- Clean agent gaseous fire suppression system shall be considered.
- Fire detection and alarm system to be provided.
- Early warning (aspiration) smoke detection system shall be provided.
- For common areas other than electrical rooms, wet sprinkler system shall be provided.
- First-aid fire extinguisher shall be provided in all areas (1 per 100 m²)
- 2-hour fire-rated doors shall be considered for rack/server hall and its associated MEP service rooms.
- The electrical and HVAC distribution shall be provided as per the fire compartments.

d. HVAC
- Operational parameters – Temperature and humidity shall be maintained to meet the requirement and shall meet the recommendation of ASHRAE.
- The system shall have chiller A (Active-1) and chiller B (Active-2) as a redundant (system N + N redundancy); both shall be equally (50%) loaded, and in the absence of one equipment, the other shall be capable of taking 100% load.

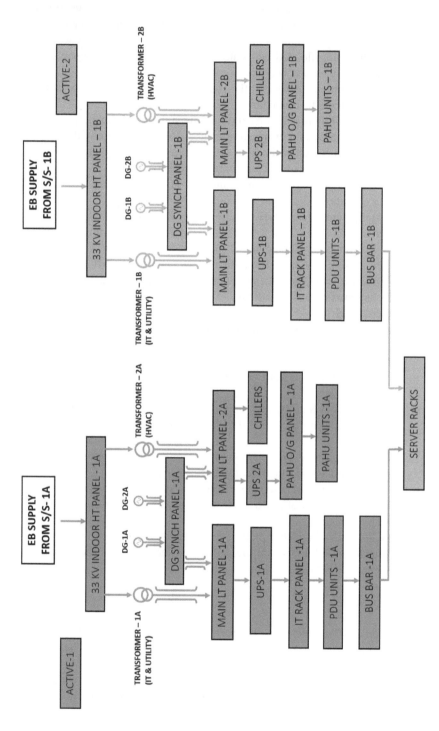

FIGURE 6.1 Tier IV data center electrical distribution.

- All the critical HVAC equipment shall be supplied from two different electric power sources as shown in Figure 6.1, and this distribution shall be through two different paths.
- In case the utility supply fails, there will be the changeover operation from utility to DG. This changeover operation will take certain amount of time in which the chillers will be without any supply that may lead to the increment of air temperature inside the data center-critical areas.
- Continuous cooling is the requirement, and to achieve that buffer tank is provided in the CHW system, which will hold the capacity of chilled water to circulate to the PAHU (precision air-handling unit) till the changeover takes place and chiller ramp up to the required capacity. CHW circulation takes place through secondary pumping system connected to the UPS power.
- The next important criteria for the Tier IV are automatic sensing and automatic containment of water leakage. This can be achieved by using the combination of leak detection systems and motorized butterfly valve (on/off).
- In case any water leak is detected in the line, the upstream motorized valve will be automatically closed, and the respective CHW path will be out of operation (entire N system) and other N system will take the operation.
- Continuous operation – HVAC system shall be available 24*365 by any means.
- Standby operation – All HVAC systems shall be connected with active power and HVAC system ready to operate at any time.

e. Fuel Leak Detection System
- Diesel generators are considered the primary source of power for Tier IV data center as per uptime institute, which requires diesel storage capacity for 12-hour continuous operation of DG sets.
- The uptime criteria also require a redundant and fault-tolerant diesel storage, and these redundant storage tanks shall feed the day tanks in two different paths.
- The fuel distribution from the diesel storage tank to the DG set shall have the following systems:
- Automatic leak detection – The leak detector sensors are placed in a fuel transfer pipe, which detects the leak and communicates to the control Panel-A (Active-1).
- Isolation of the leak detected in the distribution path – The control Panel-A (Active-1) will isolate the leaked distribution Path-A (Active-1).
- Automatic changeover to the redundant distribution path from redundant storage system – The control Panel-A (Active-1) is connected to IBMS-A (Active-1) and communicated about the leak detection. Further IBMS-A (Active-1) communicates to IBMS-B (Active-2), which in turn enables the Control Panel-B (Active-2) to automatically change over to the redundant distribution Path-B (Active-2).

f. ELV

- PLC and IO shall be selected dedicatedly for electrical, fuel station, PAHU and AHUs, chiller and primary pumps, and secondary pumps.
- Each PLC shall be equipped with 2 Nos. Ethernet ports and 2 Nos. of power supply units as a power supply redundancy.
- Each IO rack shall be equipped with 2 Nos. of power supply units as a power supply redundancy and with fiber interface unit consisting of 2 Nos. of fiber ports.

The modular UPS technology gives an advantage of switching on and off the modules as per the actual load requirements and can result in up to 5% energy savings.

SBDC uses SBMS (Smart Building Management System). In recent times, the advancements in IoT and smart devices led to the emergence of different sensors and enhanced sensing techniques with which it has become easier to keep continuous monitoring on individual parameters. Increased network speed has speeded up the data collection and processing, which has resulted in quick decision making and prompt physical response.

In SBMS, there are numerous numbers of sensor, which senses the data in almost real time. Approximately 40% of the total energy consumed is for cooling the IT equipment. PAHU /CRAC (computer room air conditioner) is the HVAC equipment, which provides air condition to the data center hall where the server racks are kept. The temperature of the data center hall is continuously monitored by various temperature sensors, and accordingly, the air flow of the PAHU/CRAC is adjusted in real time.

6.3   CONCLUSION

The smart buildings are the building blocks in a modern trend of building the smart cities. A smart city will consist of a number of smart buildings, and a number of smart buildings will generate tremendous amount of data continuously. To handle and store this huge data, SBDC can be the correct solution. The data center is the heart of the building; Tier IV topology is proposed in a SBDC to provide an efficient and more secure solution. This paper describes the design criteria to build a Tier IV data center focused on MEP services. The paper also contains the standard recommendation and criteria followed by Uptime Institute. Tier IV is a completely fault-tolerant and redundant configuration, which can be the perfect match to the SBDC. SBDC has an extra added advantage of approximately 15%–20% energy savings and efficient data storage with a very effective utilization of energy.

REFERENCES

[1] Hassan R., Bhargav K., Nishitha N. (2020). SBDC: SMART BUILDING DATA CENTER FOR IOT, EDGE, AND 5G: Spring Sim'20, May 19-May 21, 2020, Fairfax, VA, USA; ©2020 Society for Modeling & Simulation International (SCS).
[2] https://uptimeinstitute.com/tier-certification/.

7 IoT-Based Data Collection Platform for Smart Buildings

S. Chowdhury
National Institute of Technology Calicut

K.D. Saha
University of Southern California

C.M. Sarkar
MAHAGENCO

O.V. Gnana Swathika
Vellore Institute of Technology

CONTENTS

DOI: 10.1201/9781003240853-7

7.1 INTRODUCTION: BACKGROUND AND DRIVING FORCES

An Internet of Things (IoT)-based data collection platform for smart buildings is developed. A two-bus DC network is initially chosen as a test system. Fault analysis is used to analyze the normal and fault currents at the different buses. A hardware prototype is realized for a scaled-down version of the above network and IoT is incorporated through Arduino microcontroller and Node MCU (ESP8266) to collect the various data like current values, temperature, and humidity values and to visualize these data on the IoT-based Thing Speak platform. The temperature and humidity parameters are collected for a bus located at some remote location where human intervention is not possible and the corresponding relay is tripped whenever there is any abnormality in the temperature or humidity values. Three relays are used to implement manual control over the three loads from the HTML page. The concept of this small two-bus network is extended to a larger distribution network like 21 or 52 bus network using "Zonalized Data Collection" [1–9].

7.2 OBJECTIVES

The main objective is to develop an IoT platform for the collection and visualization of data, which includes various parameters, viz., current, temperature, humidity, and voltage drop of the scaled-down network designed on breadboard. Hence, a module is required that connects the grids in order to achieve remote control from almost anywhere in the world. The aim is to implement manual control over switching of loads through relays at the two buses through a self-monitored HTML page, and to visualize the voltage dip due to the presence of impedance in the first bus, manual control is implemented over a remote location bus where human intervention is not possible due to harsh weather conditions.

7.3 PROBLEM FORMULATION

The entire problem is divided into multiple modules to easily understand the main aim and objectives of the work. The first module is about choosing the test DC network, conducting short-circuit and load flow analysis to calculate the normal current and the faulted current assuming a downstream fault. The second module is the development of the scaled-down prototype network on a breadboard with one supply, three loads, three manually controlled relays, Arduino microcontroller, and Node MCU for the establishment of Wi-Fi connection required for manual relay control using HTML page. The third and the final module is the development of the Thing Speak IoT platform for the visualization of the various network parameters. Figure 7.1 shows the

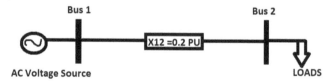

FIGURE 7.1   Two-bus test DC network.

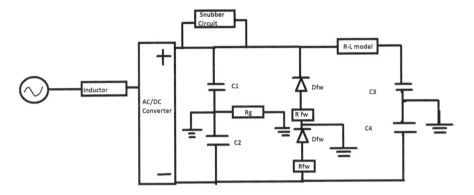

FIGURE 7.2 Equivalent network of the transmission line.

two-bus DC network, which is used as the test network. The equivalent network of a single distribution line is modeled in Laplace domain, and fault current is calculated (Figure 7.2).

7.4 DESIGN OF THE SCALED-DOWN NETWORK ON BREADBOARD

Followed by the analysis of the actual network, a scaled-down version of it is realized on a breadboard. It consists of a supply (9-V battery), which powers three loads. Each load is connected with a relay, which is controlled manually through a HTML page as an ON/OFF switch. Current sensor (ACS 712) is used to sense the current through the first LED. Temperature and humidity sensor module is used to sense the temperature and humidity of a bus at a remote location where human intervention is not possible [10]. Arduino microcontroller is used for interfacing, and Node MCU (ESP8266) is used for establishing Wi-Fi connection and serial communication between the breadboard circuit and the self-monitored HTML page.

7.5 DESIGN OF THE THING SPEAK PLATFORM FOR VISUALIZATION OF THE PARAMETERS

Thing Speak is a freely usable IoT application, which is user-friendly. It can be used to stock and recover data using different protocols over the World Wide Web or via a Local Area Network. The Thing Speak API is an interface that attends to incoming data, processes it, and generates the output readable for both humans (through well-illustrated graphs and photographs) and machines (through easily compatible codes). Thus, IoT acts as a boon for the humans as well as machines. It fastens the process of automation in industries. Here in this project, a Thing Speak platform is developed to visualize the various parameters like the current flowing through the loads, temperature, and humidity of the surroundings. Visualization of data becomes very easy through IoT by the implementation of different charts and diagrams. Therefore, it is visually attractive and is much comfortable to analyze the collected data compared to other APIs.

7.6 DESIGN OF THE IoT PLATFORM

The work basically comprises two modules. The first module is the design of the scaled-down network in breadboard, and the second module is the development of the IoT platform for data visualization. Both the modules follow different design approach. The calculations for the test network give fault current of the order of kilo amperes. Based on the feasibility of the project, the values are scaled down for design of the prototype in breadboard. Based on the decided current threshold values, appropriate relay is chosen and incorporated in the network for manual control of the loads through the self-monitored HTML page. Things Speak platform is designed with a proper Arduino Code to show the temperature and humidity variation of a remote location bus where human intervention is difficult (Table 7.1).

7.7 DESIGN SPECIFICATIONS

7.8 DESIGN OF THE HARDWARE MODEL

An IoT-based data collection platform for smart building is developed. A two-bus DC network is initially chosen as a test system. Fault analysis is used to analyze the normal and fault currents at the different buses. A hardware prototype is realized for a scaled-down version of the above network, and IoT is incorporated through Arduino microcontroller and Node MCU (ESP8266) to collect the various data like current values, temperature, and humidity values and to visualize these data on the IoT-based Thing Speak platform. The temperature and humidity parameters are collected for a bus located at some remote location where human intervention is not possible and the corresponding relay is tripped whenever there is any abnormality in the temperature or humidity values. Three relays are used to implement manual control over the three loads from the HTML page. The concept of this small two-bus network is extended to a larger distribution network like 21 or 52 bus network using "Zonalized Data Collection." First, the Arduino code for the development of the HTML page is uploaded on the Arduino board. The serial monitor shows an URL with which any device (preferably smartphone) is connected. So the displayed URL is typed in any browser and the developed HTML page, as shown in Figure 5.3, is displayed on the smartphone. The HTML page consists of four switches, and it also displays the

TABLE 7.1

Specifications of the Design Components

Component Name	Specification
Current sensor ACS 712	Current rating: 20 A
Temperature and humidity sensor module DHT 11	Including resistive humidity sensing component and negative temperature coefficient (NTC) temperature testing
Navinex relay	Current rating: 7 A
	Voltage rating: 250 V AC, 150 V DC
Boat relay	Current rating: 5 A
	Voltage rating: 125 V AC, 28 V DC

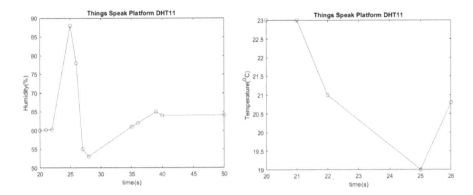

FIGURE 7.3 Plot of humidity and temperature variation in Things Speak platform.

status of the load, i.e., the LED, ON, or OFF. Based on the four switches, two for each bus, different test cases are formed. The first case is when the ON switch of bus 1 is pressed, which activates the two loads at the first bus. The second case is when the ON switch is pressed for second bus, which activates the load at the second bus. The third case is when the OFF switch of both buses is pressed, which switches off both the loads (Figure 7.3).

7.8.1 CASE 1: ON SWITCH OF BUS 1 IS PRESSED

As soon as the URL is entered in a browser of the smartphone, the HTML page is displayed on the screen with two switches for each bus, one ON and another OFF switch. The ON switch of the first bus is pressed, and with a very negligible delay of few seconds, the first relay is activated by making a sound and the two loads connected at the first bus are activated. The second load is separated from the first by a resistance. Hence, there is a voltage drop between the two loads. This drop in voltage is sensed by a voltage sensor. The first LED glows brighter than the second one, which shows that there is a voltage drop. This case is shown where it is seen that only the two LEDs at the first bus are glowing (Figure 7.4).

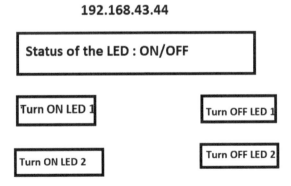

FIGURE 7.4 Developed HTML page.

7.8.2   Case 2: ON Switch of Bus 2 Is Pressed

The ON switch of the second bus is pressed, and with a very negligible delay of few seconds, the second relay is activated by making a sound and the load connected at the second bus is activated. This bus is assumed to be at a remote location where human intervention is not possible due to harsh geographical conditions. This case is shown where it is seen that only the LED connected to the second bus is glowing (Figure 7.5).

7.8.3   Case 3: OFF Switch of both Buses Is Pressed

The OFF switch of the first bus is pressed followed by that of the second bus, and with a very negligible delay of few seconds, both the relays are activated by making a sound and the load connected at both buses is deactivated. This case is shown where it is seen that only the LED connected to the second bus is glowing (Figure 7.6).

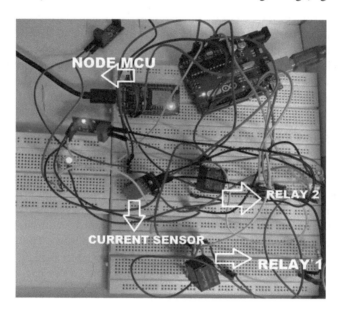

FIGURE 7.5   Case 1: Hardware setup with only first bus activated.

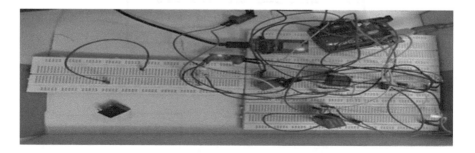

FIGURE 7.6   Case 2: Hardware setup with only second bus activated.

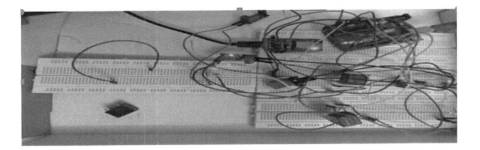

FIGURE 7.7   Case 3: Hardware setup with both buses deactivated.

7.8.4.   CASE 4: ON SWITCH OF BOTH BUSES IS PRESSED

The ON switch of the first bus is pressed followed by that of the second bus, and with a very negligible delay of few seconds, both the relays are activated by making a sound and the load connected at both the buses is activated. This case is shown where it is seen that only the LED connected to the second bus is glowing (Figure 7.7).

7.9   RESULTS AND DISCUSSIONS

The following are the results obtained:
   i. Temperature value   : 31.00°C
   ii. Humidity percentage   : 62.00%
   iii. Voltage drop   : 2.49 V (when LED ON)
                               3.32 V (when LED OFF)
   iv. Current   : 0.49 mA

Based on the results presented above, we can say that the temperature, humidity, and other parameters can be estimated for a bus located at a geographically remote location using the concept of IoT (Figure 7.8).

FIGURE 7.8   Case 4: Hardware setup with both buses activated.

7.10   CONCLUSION

The conventional power grid provides reliable and uninterrupted power throughout the year to our households and industries. But when the grid is threatened by disasters and security issues, the following blackouts cause a lot of havoc. Hence, organizations and utilities are striving hard together to construct strong and resilient power systems known as microgrids. Microgrids can operate either as part of the conventional power grid or independently (or both), revolutionizing the way the energy resources are managed.

A small prototype of a high-voltage DC network is realized on a breadboard and is tested with real-time data. The circuit consists of a supply of 5 V from the Arduino board, three LEDs (as loads), and Node MCU (ESP 8266) for enabling Wi-Fi connection. A HTML page is developed, which consists of four switches for turning on and off the loads. Current sensor is incorporated in the first bus, which senses the current once the load is switched on from the HTML page. The first bus consists of two loads. A resistance is placed in between the two loads to show a voltage drop. The voltage drop is being sensed by the voltage sensor module placed in between the two loads in the first bus. The first bus is at normal conditions and is controlled from the HTML page. The second bus is assumed to be at a remote location where there are very harsh temperature or humidity conditions. Whenever any abnormality in temperature is observed, the load is switched off from the manually controlled HTML page. A voltage drop of about 2.5 V is observed using a voltage sensor, and current is sensed using ACS 712 sensor. Any abnormalities above the threshold values are treated as a fault, and accordingly, the loads are turned off.

7.11   FUTURE SCOPE

The work is done keeping in mind a scaled-down prototype of the test two bus DC network. The normal and fault currents for the hardware setup are scaled down accordingly, and the entire analysis is done for two buses consisting of a single common supply and three loads (three LEDs). The project has very wide scope; for example, it is implemented to control a bus which is at a very remote location where human intervention is not possible due to harsh weather or some other unforeseen circumstances. Control to such a remote location is enabled from any part of the world with just a smartphone, which has an internet connection.

As soon as the temperature or humidity of that remote location goes above certain normal threshold manually, the loads are switched off from the HTML page. Moreover, as the temperature or humidity reaches a normal value, the load is again switched on for normal operation. Here, a small load like a LED is used, but this setup is used for larger loads also; accordingly, the ratings of the devices need to be changed.

In future for further analysis of larger networks consisting of 21 buses or 52 buses, the entire system is subdivided into various zones and zonal analysis is carried out. This concept of dividing the network into zones is termed as *"Zonalized Data Collection."*

REFERENCES

[1] Swathika, OVG, and KTMU Hemapala. IOT based energy management system for standalone PV systems. *Journal of Electrical Engineering & Technology* 14, 5, 2019, 1811–1821.

[2] Soj, RP, S Saha, and OVG Swathika. IoT-based energy management system with data logging capability. *Proceedings of International Conference on Sustainable Computing in Science, Technology and Management (SUSCOM), Amity University Rajasthan, Jaipur-India*, 2019.

[3] Gupta, Y, et al., IOT based energy management system with load sharing and source management features. *2017 4th IEEE Uttar Pradesh Section International Conference on Electrical, Computer and Electronics (UPCON)*, IEEE, 2017.

[4] Ananthakrishanan, V, et al., GSM based energy management system. *International Journal of Pure and Applied Mathematics* 118, 24, 2018.

[5] Swathika, OVG, et al., IoT-based energy management system with data logging capability. In *Advances in Smart Grid Technology*, Springer, Singapore, 2021, pp. 547–555.

[6] Majee, A, and OVG Swathika. IoT based reconfiguration of microgrids through an automated central protection centre. *2017 International Conference on Power and Embedded Drive Control (ICPEDC)*, IEEE, 2017.

[7] Singh, A, et al., Arduino based home automation control powered by photovoltaic cells. *2018 Second International Conference on Computing Methodologies and Communication (ICCMC)*, IEEE, 2018.

[8] Majee, A, M Bhatia, and OVG Swathika. IoT based microgrid automation for optimizing energy usage and controllability. *2018 Second International Conference on Electronics, Communication and Aerospace Technology (ICECA)*, IEEE, 2018.

[9] Swathika, OVG, and S Hemamalini. Prims-aided dijkstra algorithm for adaptive protection in microgrids. *IEEE Journal of Emerging and Selected Topics in Power Electronics* 4, 4, 2016, 1279–1286.

[10] Swathika, OVG, and KTMU Hemapala. IOT-based adaptive protection of microgrid. *International Conference on Artificial Intelligence, Smart Grid and Smart City Applications*, Springer, Cham, 2019.

8 Sensor Data Validation for IoT Applications

Siddharth Pandya, C. Sai Aditya, and
Abraham Sudharson Ponraj
Vellore Institute of Technology

CONTENTS

8.1 INTRODUCTION

The world we live in today uses technology to make our lives easier and smarter. Sensors play an essential role in this case. There has been a vast deployment of various types and sizes of IoT (Internet of Things) devices at different parts of the modern ecosystem comprising urban areas with buildings with offices and houses, lawns, and even city-wide rural agricultural-level farms [1]. The huge number of sensor data streams obtained by these systems causes the issue of scalability, which is one of the key challenges in this rapidly growing field. However, just by merely connecting the large number of sensors in a mesh network followed by collecting their data from input streams and storing, only just retrieving these stored sensor data is not a wise use of this data. Thus, it provides no benefits that can be considered helpful. Instead, when this stored data in itself can be processed and analyzed smartly to predict some other important decisions, that makes up for the unique core features of IoT devices and some of the helpful services provided by its ecosystem. Such well-processed and analyzed data can be used in various creative aspects and activities like making decisions to control things, generate alerts, or combine two or more sets of processed and analyzed data to obtain some even more useful data. Since almost all the IoT applications in the present era rely on sensors in the field to get some reliable data from them, it becomes an important step to get the sensor data validated and further provide that clean, correct data for further decisions and analysis. Sensors provide a fantastic connection to the physical world, but extracting usable data is not simple. Many first-time IoT users are unprepared for how messy a sensor's data can be (Dirty Data) [2].

DOI: 10.1201/9781003240853-8

Moreover, the thing that should be considered first and foremost is that the data gathered/collected through various sources must be clean and correct and useful for the case, making this step vital for data validation. Considering the case that if this data comes from a considerable number of sources, maybe millions of sources, each streaming billions of data stream, then every data stream must be clean and corrected (if found with any errors) before using it in any data analysis gets some desired results. Furthermore, if this input data is not validated, it will lead to incorrect and many times skewed results, leading to considerable impact. Therefore, data validation is not only necessary but vital for the sensors used in IoT applications. Hence, in conclusion, the research work's objective is to build an environment for data validation of sensors in various IoT applications.

The further sections in this research work have been divided as follows: Section 8.2 discusses various validation methods based on research done in various research articles, which summarize the positives and negatives of their methods, leading to implementing the proposed system on data validation. Section 8.3 summarizes the proposed system design to give an overview of how the core ideas are implemented to solve the problem. Then, the detailed implementation of the proposed system is discussed in Section 8.4. Section 8.5 illustrates the results and outputs obtained, and Section 8.6 gives the conclusion.

8.2 LITERATURE REVIEW

Input data points from different sources are analyzed by an IoT framework, which are used in practical decision making. One of the key requirements that an IoT application demands is quality assurance. Quality assurance can make the system more expensive. Once the data is on the cloud, the importance for data accuracy is underlined. In the cloud data collection process, an input data validation filter layer could be added to improve the system efficiency. This section will conduct potential validations and eliminate invalid data points, document the results, and store them. This was focused on the proposed system. Training of prediction models is done using these findings (ARIMA in this paper) for sensor failure, data interpolation, and sensor data validation [3].

The behavior of the process in standard operation condition is monitored to validate the sensor data. The dynamics of the values can be modeled, i.e., the variable level can be determined. A physical parameter is influenced by external factors; for instance, a room temperature tends to rapidly change by opening the window in a room. Therefore, distinguish two regular operation states: (i) when the process is in a stationary state and (ii) when an external factor influences the process [4].

A general and efficient approach to data distribution is another way of validating data that does not require foreknowledge of the input distribution. On the basis of this methodology, an efficient technique is used for the quantification of distributed deviations in the sensor network. The aim is for all sensor inputs in a sliding window to recognize values that have a very few nearby adjacent. This problem is particularly crucial in the sensor network setting because it can identify faulty sensors and filter false reports from various sensors [5].

TABLE 8.1
Sensor Faults

Fault	Sensor Fault Type
1	A data point that deviates significantly from the expected temporal trend of the data
2	A rate of change much more significant than expected over a short/longer period, which may or not return to normal afterward
3	A series of data points having zero or almost zero variations for some time higher than expected
4	Sensor data exhibiting an unexpectedly high amount of variation over the temporal domain
5	Sensor data may have offset or had a different gain from the ground-truth values

Data validation is done to ensure that sensor faults do not occur. These sensor faults can happen in several ways. Table 8.1 shows some of the faults, which most likely occur in an IoT application [6,7].

Fall curve is a non-rule-based approach that is used for data validation. On trying to implement it, the problem was that the hardware implementation circuit required an ADC component. Arduino was used for this purpose. However, Arduino has an independent power source, so fall curve could not be obtained. It was challenging to obtain analog circuitry capacitive value for different sensors, which determines fall curve [8].

8.3 PROPOSED SYSTEM

In this research work, a technique is presented that detects sensor events that are abnormal such as sudden voltage fluctuations in the power input of IoT hub and physical damages in the IoT hub box. These abnormal detections are obtained from the behavior of the procedure, which is being monitored based on procedure's theoretical understanding.

The main benefit of the above procedural understanding method is that it is built upon or is derived from the theoretical and structural works that are so well established that they deliver a proper methodical and practical approach that requires low computational requirements. Live temperature data from an IoT sensor is collected, on the one hand. On the other hand, a machine learning algorithm has been created, which predicts the present-day data using a model trained with previous 20-year temperature data.

Both are compared. If the data received is faulty, then the system corrects and displays it to the user; else, the original temperature from the IoT sensor is displayed. Our web application then can be deployed so that anyone can access it globally using cloud [9]. This will eliminate the need for a centralized control unit. The user can see the real-time output in the form of a webs service where the data is being read and corrected [10]. This collected dataset can be used for further analysis and research. The structure of the proposed system can be seen in Figure 8.1.

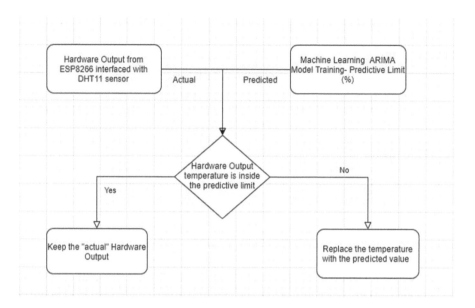

FIGURE 8.1   Proposed system design.

8.4   IMPLEMENTATION

The proposed system incorporates the combined framework for the hardware and the software part. The hardware part includes the IoT hubs, including the temperature sensor (DHT11) and the ESP8266 board, to capture the sensor readings and then process it to load the reading onto the server in real time [11]. Figure 8.2 shows the hardware setup with the microcontroller and the sensor interfaced.

The code to capture the reading from the sensors is embedded onto the board using Arduino IDE, which on running displays the real-time temperature on the serial monitor [12]. These readings are then displayed on a local host based

FIGURE 8.2   Hardware setup.

FIGURE 8.3 Temperature output.

on the IP address of the web server. This will show the live data of the sensor as shown in Figure 8.3.

Next is the creation of the machine learning model, i.e., autoregressive integrated moving average (ARIMA). The weak point of the autoregressive moving average (ARMA) model can be overcome by including an integral parameter to the model. That is, differentiate (d) the ARMA (p, q)-based time-series model at the beginning, and then they are integrated to produce the final predicted values. The ARIMA (p, d, q) model is neatly described in Equation (8.1) with the help of the backshift operators that can predict the future value [13,14].

$$Y_t = \alpha + \beta_1 Y_{t-1} + \beta_2 Y_{t-2} + \ldots + \beta_p Y_{t-p} \in_t + \varphi_1 \in_{t-1} + \varphi_2 \in_{t-2} + \ldots + \varphi_q \in_{t-q} \quad (8.1)$$

The value that is predicted using this model is the sum of a constant (α) and the linear combination of lags of target variable (up to p lags) and lagged forecast errors (up to q lags).

In the broad domain of statistics and econometrics, going into the specific topic of time-series analysis, an ARMA model is the part of the ARIMA. Considering both these models, they can be used to fit time-series data to obtain two of the following things: Understand the data better and predict future points in the time series [13].

Jupyter Notebook is used for running the code of the ARIMA model. The model is trained using raw data. This raw data is the temperature and humidity data of Chennai from 1995 to 2019. After this, the results of the model are compiled, and the predicted vs. expected data is shown as the output [15].

In the model results below, the p, d, and q values have been taken as 1,1,1. On one time, differencing was required to make the series stationary. This specific ARIMA model is the combination of autoregression model, which is of first order, and a moving average model, which is also of first order. Then using the Kalman filter, the

```
                                    ARIMA Model Results
==============================================================================
Dep. Variable:          D.Temperature   No. Observations:               8957
Model:                  ARIMA(1, 1, 1)  Log Likelihood            -11835.039
Method:                        css-mle  S.D. of innovations            0.907
Date:               Mon, 09 Mar 2020    AIC                        23678.078
Time:                      23:27:15     BIC                        23706.479
Sample:                           1     HQIC                       23687.744
```

FIGURE 8.4   ARIMA model results.

exact probability is computed, which is determined by the starting values obtained by maximizing the conditional sum of squares likelihood [16]. The result can be seen in Figure 8.4, which shows the total number of observations in the dataset, which is 8957.

The standard deviation is 0.907. The following results are tabulated in the form of three terms of ARIMA model, which are as follows: constant term (α), autoregressive term (lags of y), and the moving average term (lagged forecast errors). This table displays all the statistics regarding the independent variables. Considering the column with "p" values in it, we observe that based on the coefficients of all the three terms, the constant plays a slightly more important part in the equation. Also, the AR and MA terms help in capturing the previous trend in the graph of temperature readings in order to predict the future trends (Figure 8.5).

The next step towards the implementation is setting up a serial communication between Arduino IDE and Python script. The connection does the work of collecting temperature from NodeMCU using DHT11 sensor and serially transferring it to the python platform. The final step involved the creation of a web application for displaying the validated real-time data input [11].

Streamlit library is used for making web applications using Python. Streamlit then displays a sectioned web application where it shows moments' real-time temperature, the machine learning model predictions, and the data validations for the sensor's

```
------------------------------------------------------------------------------
========
                           coef    std err          z      P>|z|      [0.025
0.975]
------------------------------------------------------------------------------
--------
const                    0.0005      0.003      0.171      0.864      -0.005
0.006
ar.L1.D.Temperature      0.5840      0.013     43.452      0.000       0.558
0.610
ma.L1.D.Temperature     -0.8691      0.008   -112.373      0.000      -0.884
-0.854
                                    Roots
==============================================================================
                 Real          Imaginary          Modulus          Frequency
------------------------------------------------------------------------------
AR.1           1.7124          +0.0000j           1.7124             0.0000
MA.1           1.1506          +0.0000j           1.1506             0.0000
------------------------------------------------------------------------------
```

FIGURE 8.5   ARIMA model coefficients.

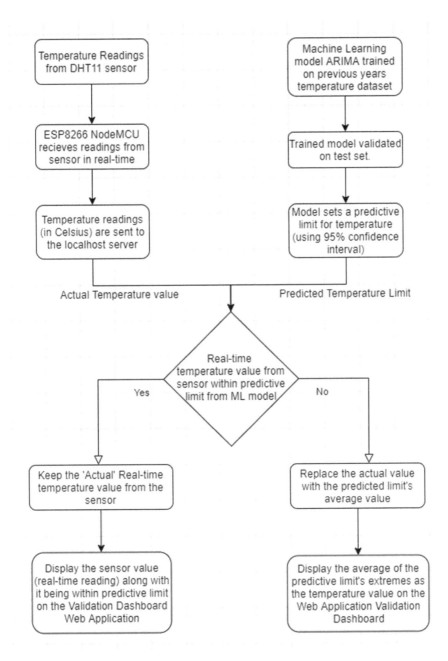

FIGURE 8.6   Detailed system design.

readings with respect to the model onto a web application [17,18]. This complete procedure can be viewed in a concise manner in the block diagram in Figure 8.6.

8.5   RESULT AND DISCUSSION

In this paper, the web application shows the temperature readings which are being recorded in real time by the sensor output of the IoT hubs, which are deployed at various locations, for example, at major landmarks in a smart city scenario.

The server for the web application does the process of collecting the recorded temperatures, analyzes their pattern or trend, then validates the reading for its correctness, and finally displays the whole process on the web page, which can further be accessed at various locations.

The web app has a sectioned display, which starts as follows:

1. The first part shows the real-time data received at the point of time when the web app is being accessed at any location and tabulates it with respective timestamp as mentioned above, which is then followed by a real-time plot for that point of time.
2. The second part is where the machine learning models' predictions are tabulated (expected vs. predicted). The model is trained on a previous 20-year dataset for the temperatures of the city.
3. The third and final part is where the validation is displayed, in which the real-time readings' average value is taken, compared with the machine learning models' predictive limit, and if it is within the limit, then it validates the real-time reading as correct and is displayed as final temperature. If the real-time average is out of the predicted limit, then the web app outputs the limit and its median value as the temperature value for that instance of time instead of the reading which came in through the sensor in IoT hub box.

The webpage output displays a set of ten readings along with timestamp. A graph of the real-time plot is depicted. The predicted vs. expected values for the ten readings are shown, and finally, it shows whether this dataset requires data validation. Figures 8.7–8.9 show the web pages for all possible cases.

The output is represented by three sets of web application outputs, which are as follows:

1. Figure 8.7 is for when the temperature lies within the predictive limit. Here, the real-time temperature has been obtained as 24.83°C from the hardware setup. The predictive limit produced by the model is between 24.38°C and 27.73°C. Since the real-time data lies within this limit, it is the right output.
2. Figure 8.8 occurs when output is below the predictive limit. The real-time value obtained is 17.75°C, which is below the predictive limit. Therefore, the expected value calculated by the model is 26.067°, which is displayed as the correct value instead.

Temperature Machine

Real-Time Temperature Readings

	Date	Temperature
0	2020-05-01	27.6000
1	2020-05-01	27.6000
2	2020-05-01	27.8000
3	2020-05-01	27.8000
4	2020-05-01	27.8000
5	2020-05-01	27.9000
6	2020-05-01	27.9000
7	2020-05-01	27.9000
8	2020-05-01	28
9	2020-05-01	28

Conclusions from Real-Time temperature table and plot

The real-time average temperture from the IoT sensor records **24.8300 degree Celcius**

Model Predictions

The following are the forecasted predictions and actual readings:

	Predicted	Expected
0	30.2271	29.8333
1	29.9245	31
2	30.5501	28.6111
3	29.8195	29.5556
4	29.8533	29.7778
5	29.4533	30.8444
6	30.4984	31.2222
7	31.5492	31
8	30.7283	31.6667
9	31.4338	31.6667
10	31.2878	31.6667

The predicted temperature by model is **26.0637 degree Celcius**

The model-predicted range in which the temperature from IoT sensor can lie is between 24.390 degree Celcius and 27.7375 degree Celcius

Data Validation for IoT Sensor

The real-time value from sensor which is **24.830000 degree Celcius** lies well within the model predicted range i.e. Between **24.389993 degree Celcius** and **27.737490 degree Celcius**

Hence the sensor is working correctly.

And the correct present temperature is **24.8300 degree Celcius** as recorded by the IoT sensor.

Made with Streamlit

FIGURE 8.7 Web application preview for the correct predictive limit.

Temperature Machine

Real-Time Temperature Readings

	Date	Temperature
0	2020-05-01	25.2000
1	2020-05-01	24.8000
2	2020-05-01	23.7000
3	2020-05-01	22.4000
4	2020-05-01	21.2000
5	2020-05-01	19.9000
6	2020-05-01	18.9000
7	2020-05-01	18
8	2020-05-01	17.1000
9	2020-05-01	16.3000

Real-time plot

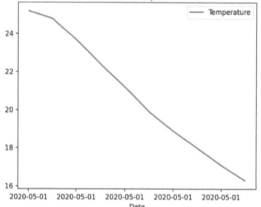

Conclusions from Real-Time temperature table and plot

The real-time average temperture from the IoT sensor records **17.7500 degree Celcius**

Data Validation for IoT Sensor

The real time value from sensor may be incorrect, as it is out of model predicted range

Hence the sensor is not working correctly

So the correct present temperature should be **26.0637 degree Celcius** and not the temperature recorded by IoT sensor (i.e. *17.7500 degree Celcius*)

Made with Streamlit

FIGURE 8.8 Web application preview for the below predictive limit.

Temperature Machine

Real-Time Temperature Readings

	Date	Temperature
0	2020-05-01	30.1000
1	2020-05-01	30.2000
2	2020-05-01	30.2000
3	2020-05-01	31
4	2020-05-01	33.3000
5	2020-05-01	36.8000
6	2020-05-01	39.8000
7	2020-05-01	42.4000
8	2020-05-01	44.3000
9	2020-05-01	45.5000

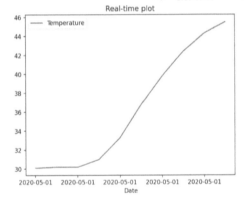

Conclusions from Real-Time temperature table and plot

The real-time average temperture from the IoT sensor records **33.3600 degree Celcius**

Data Validation for IoT Sensor

The real time value from sensor may be incorrect, as it is out of model predicted range

Hence the sensor is not working correctly

So the correct present temperature should be **26.0637 degree Celcius** and not the temperature recorded by IoT sensor (i.e. 33.3600 degree Celcius)

FIGURE 8.9 Web application preview for the above predictive limit.

3. Figure 8.9 is for the temperature above the predictive limit. The real-time value obtained is 33.36°C, which is below the predictive limit. Therefore, the expected value calculated by the model is 26.067°, which is displayed as the correct value instead.

So, it is possible to observe that the first real-time temperature reading received from the sensor is well within the range predicted by the machine learning model. Then, in the second output, we can observe that the real-time temperature goes way below the predictive limit by the model. That is, way below what the actual temperature should be. So, the model validates the temperature reading, and instead of displaying the wrong one, it corrects it and then displays the model's predicted temperature for this scenario. Finally, in the third set of real-time readings output, temperature goes way above the predictive limit. So, it also validates it and displays the correct temperature.

Other than the web application output, the machine learning model ARIMA, a time-series model, runs in the back-end, i.e., at the local host server side, which computes the predicted vs. expected values for the temperature readings. In Figure 8.10, x-axis represents the number of observations in the test dataset, which is around 3,000, and y-axis represents the temperature in degree Celsius [19]. The near-overlap in the two sets of values (predicted and expected) can also be seen in the figure, which depicts that the model has fitted to the training data and how well the trained model has predicted and forecasted on the training data.

MSE is the mean standard error. MAE is the mean absolute error. The coefficient of determination is 0.896, which means that the model is 89.6% accurate.

```
Test Mean Squared Error: 0.712
Test Mean Absolute Error: 0.616
Test Coefficient of Determination: 0.896
```

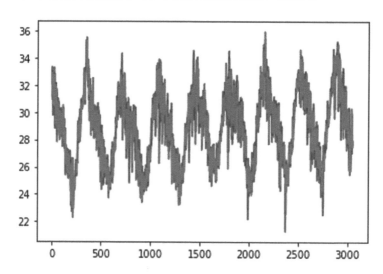

FIGURE 8.10 Overlap of predicted and expected graphs.

8.6 CONCLUSION

Sensor validation thus plays an important role in each and every condition where a particular IoT device is used to log various data using its sensors. And as sensors operate on power, so they are prone to malfunction and read in faulty values, which at a later stage could cause issue in analyzing the data streams when they have to be used for further predictions or decisions in various other fields. So, in order to have an error-free data collection in real time, the system proposed come into use here. The sensor data validation system combining the use of ESP8266 NodeMCU with DHT11 sensor with the help of a pre-trained ARIMA model is helpful in making decisions whether the data read is faulty or not and hence correcting it if it is faulty. This whole system is a very lightweight system, and hence in real time, it can function well in accordance with all its components.

Finally, the web application can be a very helpful dashboard to view and access the faulty sensors and the real-time corrections which are being made if data collected is incorrect as predicted by the machine learning model. The output for all possible scenarios was successfully obtained.

In a future reference, this dashboard-based web application can be deployed onto the cloud, which in turn can then be accessed by anyone in different cities. It could also aid the user to gather the data collected for any other analysis or decision making.

REFERENCES

[1] Chacko, V., & Bharati, V. (2017, June). Data validation and sensor life prediction layer on cloud for IoT. In *2017 IEEE International Conference on Internet of Things (iThings) and IEEE Green Computing and Communications (GreenCom) and IEEE Cyber, Physical and Social Computing (CPSCom) and IEEE Smart Data (SmartData)* (pp. 906–909). IEEE.

[2] Sándor, H., Genge, B., & Szántó, Z. (2017, September). Sensor data validation and abnormal behavior detection in the internet of things. In *2017 16th RoEduNet Conference: Networking in Education and Research (RoEduNet)* (pp. 1–5). IEEE.

[3] Ni, K., Ramanathan, N., Chehade, M. N. H., Balzano, L., Nair, S., Zahedi, S., ... & Srivastava, M. (2009). Sensor network data fault types. *ACM Transactions on Sensor Networks (TOSN)*, 5(3), 1–29.

[4] Subramaniam, S., Palpanas, T., Papadopoulos, D., Kalogeraki, V., & Gunopulos, D. (2006, September). Online outlier detection in sensor data using non-parametric models. In *VLDB* (Vol. 6, pp. 187–198).

[5] Chakraborty, T., Nambi, A. U., Chandra, R., Sharma, R., Swaminathan, M., Kapetanovic, Z., & Appavoo, J. (2018, November). Fall-curve: A novel primitive for IoT Fault Detection and Isolation. In *Proceedings of the 16th ACM Conference on Embedded Networked Sensor Systems* (pp. 95–107).

[6] Borgia, E. (2014). The Internet of Things vision: Key features, applications and open issues. *Computer Communications*, 54, 1–31.

[7] Chatzigiannakis, V., & Papavassiliou, S. (2007). Diagnosing anomalies and identifying faulty nodes in sensor networks. *IEEE Sensors Journal*, 7(5), 637–645.

[8] Deshpande, A., Guestrin, C., Madden, S. R., Hellerstein, J. M., & Hong, W. (2004, August). Model-driven data acquisition in sensor networks. In *Proceedings of the Thirtieth International Conference on Very large Data Bases* (Vol. 30, pp. 588–599).

[9] Sharma, A. B., Golubchik, L., & Govindan, R. (2010). Sensor faults: Detection methods and prevalence in real-world datasets. *ACM Transactions on Sensor Networks (TOSN)*, 6(3), 1–39.

[10] Puttagunta, V., & Kalpakis, K. (2002, June). Adaptive Methods for Activity Monitoring of Streaming Data. In *ICMLA* (Vol. 2, pp. 197–203).

[11] Mehrmolaei, S., & Keyvanpour, M. R. (2016, April). Time series forecasting using improved ARIMA. In *2016 Artificial Intelligence and Robotics (IRANOPEN)* (pp. 92–97). IEEE.

[12] Murat, M., Malinowska, I., Gos, M., & Krzyszczak, J. (2018). Forecasting daily meteorological time series using ARIMA and regression models. *International Agrophysics*, 32(2).

[13] Podolskiy, V., Jindal, A., Gerndt, M., & Oleynik, Y. (2018, September). Forecasting models for self- adaptive cloud applications: A comparative study. In *2018 IEEE 12th International Conference on Self-Adaptive and Self-Organizing Systems (SASO)* (pp. 40–49). IEEE.

[14] Farhath, Z. A., Arputhamary, B., & Arockiam, L. (2016). A Survey on ARIMA Forecasting Using Time Series Model. *International Journal of Computer Science and Mobile Computing*, 5, 104–109.

[15] Alsharif, M. H., Younes, M. K., & Kim, J. (2019). Time series ARIMA model for prediction of daily and monthly average global solar radiation: The case study of Seoul, South Korea. *Symmetry*, 11(2), 240.

[16] DHT11 Temperature and Humidity Sensor. DHT11 Technical Data Sheet. https://osepp. com/wp-content/uploads/2015/04/DHT11-Technical-Data-Sheet-Translated-Version.pdf

[17] Streamlit Documentation. https://docs.streamlit.io/

[18] ARIMA model Forecasting. https://www.machinelearningplus.com/time-series/ arima-model-time-series-forecasting-python/

[19] Deploy Streamlit-based machine learning model. https://towardsdatascience.com/ streamlit-deploy-a-machine-learning-model-without-learning-any-web-framework- e8fb86079c61

9 Energy Economic Appliances' Scheduling for Smart Home Environment

K Jamuna, D Suganthi, and S Angalaeswari
School of Electrical Engineering, Vellore
Institute of Technology

CONTENTS

9.1 INTRODUCTION

The home appliances used in olden days do not require electrical power, which is operated by man power and human efforts. By the advent of the electricity and for the human comfortability, many electrical appliances are entered into our home. These appliances are operated and controlled by the human. During the constant energy billing, the electricity bill was not heavy. By the increase in the use of electrical appliances, the energy need of the individual is growing day by day. In the last few decades, energy requirement has been doubled for an individual home itself. Due to that, the generation also needs to be improved. Hence, the deficit of natural resources and global warming give a warning to the electrical engineers for taking the necessary action towards the clean, economic, and sustainable energy. In addition, they need to develop the suitable methods for providing efficient and effective usage of appliances.

Many appliances have been built with intelligent thinking with IoT techniques. Hence, these appliances are named as smart appliances. These appliances are capable to operate according to the real-world situation, and their data could be sent to the cloud or local storage. Schmidt explained the above concepts with mobiles as the communication medium between real world and appliances [1].

DOI: 10.1201/9781003240853-9

9.2 APPLIANCES' CATEGORIZATION

Basically, these appliances are classified as controllable and non-controllable devices. The controllable devices/appliances are able to schedule their operating hours, whereas the non-controllable devices' scheduling is not possible [2]. The entertainment appliances like TV, computer, and sound systems are considered as non-controllable devices. Household appliances can be grouped under six categories according to Beaudin and Zareipour [3], namely, uncontrollable loads, curtailable loads, uninterruptible loads, interruptible loads, regulating loads, and energy storage. The curtailable loads (e.g., artificial indoor light illumination) are able to adjust their operation without affecting the residents' comfort. The uninterruptible loads are to be operated to complete their function, such as dish washer, cloth dryer. In contrast, the interruptible loads can be interrupted any time in between, for example, the charging of the electric vehicle and any rechargeable devices. The heating, ventilating, air-conditioning devices are operated based on the given reference, and they are named as regulated loads. Batteries and other storage devices fall under the category of energy storage devices.

Each appliance has certain minimum operating time, and rough energy consumption is also listed in Table 9.1 [4], which is based on the middle-class home appliances. The types of loads are also mentioned. The shiftable load means that they could operate anytime of the day. Among the appliances, the refrigerator is a device that needs to be powered continuously throughout the day. There are other electrical

TABLE 9.1
Home Appliances Approximate Power Consumption and Approximate Operating Time

Sl. No.	Name of the Appliances	Power Consumption(kW)	Operating Time (minutes)	Type of Load
1	Dish washer	0.6–1.2	105	Interruptible and shiftable
2	Cloth washer or washing m/c	0.52–0.65	45	Interruptible and shiftable
3	Refrigerator	0.37	24 hours a day	Continuous and non-shiftable
4	Air conditioner	0.25–2.75	Based on weather condition	Non-shiftable
5	Micro-oven	0.75–2.35	60	Shiftable
6	Electric vehicle	3	180	Interruptible and shiftable
7	Electric motor	0.746	30	Interruptible and shiftable
8	Compressor	0.746	30	Interruptible and shiftable
9	Induction stove	1.5	30	Shiftable
10	Electric heater	1	30	Interruptible and shiftable

devices like lamps, fans, laptops, and mobile charges. These devices' power consumption is much lesser than the major home appliances. Hence, those devices are not listed even they could operate with IoT.

9.3 SYSTEM DESCRIPTION

Present/future home is filled with many smart appliances, such as dish washer, cloth washer, etc. These appliances need electrical power to operate. Hence, the home power consumption will be raised that increases the energy bill. The imported energy price also varies according to the public demand. In this work, a program is developed to schedule the appliances in order to optimize the home energy consumption by considering their own power production also. It is assumed that the house is installed with solar power plant or any other renewable resources. If the power generated at the home exceeds their consumption, then it is exported to the grid. Else, the power is to be imported and utilized. The typical modeled system is shown in Figure 9.1.

Nowadays, the home appliances are intelligent enough to operate without human efforts, since many people are working and tough to operate the devices within their available time at home. The power tariff also changes time to time. Hence, it is

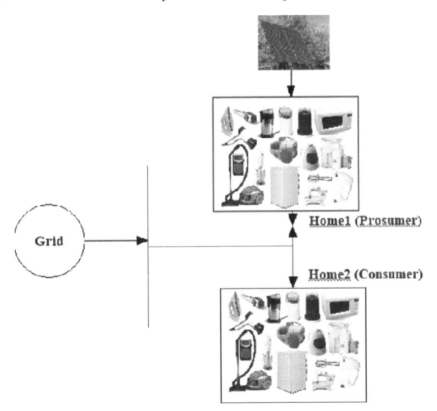

FIGURE 9.1 Typical power network connected to the utilities (Home 1 is denoted as prosumer, can import and export power, and Home 1 is denoted as consumer, can import only).

necessary to export the power when the peak demand is at the grid side and import the power to operate the appliances to reduce the energy bill.

Problem Definition: The main objective is to schedule the home appliances in such a way to minimize the home power consumption by incorporating their own power generation too. The home might be equipped with solar panel and supports their energy needs and excess power may be exported at certain periods. The main objective could be met based on different situations.

Case Study 1:

Minimize: Energy consumption of the house $\sum T * P * X$

The assumption is made in that the grid is able to supply any power demand, where T is the tariff, P is the power consumption, and X is the variable, which denotes the appliances available at the particular time.

Case Study 2:

Minimize: Energy consumption of the house

Subject to: $\sum_{i=1}^{NH} P_i \leq P_{max}$

P_i is the power consumption of the ith house. NH is the number of houses considered, and P_{max} is the maximum power demand.

In this case, the maximum power demand is fixed. According to that, the houses plan their appliances to be switched on. Based on the above, many case studies could be framed according to the situation and need, which is mentioned in Table 9.2.

TABLE 9.2

Objectives and Constraints of the Problem Statement

Case Study	Objective Function	Subject to the Constraints
1	**Minimize**: Energy consumption of the house	NIL
2	**Minimize**: Energy consumption of the house	**Subject to:** $\sum_{i=1}^{NH} P_i \leq P_{max}$
3	**Minimize**: Net energy consumption of the house = Energy consumption of the house - Solar energy produced in that house	**Subject to:** $\sum_{i=1}^{NH} P_i \leq P_{max}$
4	**Minimize**: Net energy consumption of the house = Energy consumption of the house - Solar energy produced in that house	**Subject to:** $\sum_{i=1}^{NH} P_i \leq P_{max}$ Appliance sequences should be in a particular fashion
5	Case study 4 + based on the climatic conditions	

9.4 SIMULATION RESULTS

The assumption is made that the solar panel is installed in the smart home and the different tariff patterns are assumed for weekdays and weekends. Based on the above assumptions, many case studies are performed under different climatic conditions and tariff rates. These case studies support the optimal appliance scheduling in smart home environment for different circumferences economically. In this case, the appliance is operated for 1-hour time duration and the number of appliances is considered as ten. The problem is solved using genetic algorithm, and the assumed parameters are the number of populations: 200; random crossover; and mutation probability = 0.04. The population matrix is the number of appliances by the number of slots/day of the defined problem. Here, the single population matrix size is [10×24].

Case Study A: Minimize the energy bill based on the weekday tariff. The optimized energy bill of the scheduled appliances over the number of iterations is shown in Figure 9.2. The energy bill is Rs.11.46/- when the cloth washer operated for the multiple slots.

Case Study B: Minimize the energy bill based on the weekend tariff. The optimized energy bill of the scheduled appliances over the number of iterations is shown in Figure 9.3. The energy bill is Rs.19.46/- when the cloth washer operated for the multiple slots.

Case Study C: Minimize the energy bill based on the weekday tariff with considerable solar input. The lowest energy bill is Rs.10.55/- when the solar input is also considered, which is shown in Figure 9.4, and the appliances are similar to the case study A. Here, the export power is not valued for price. Also, the energy bill based on the weekend tariff with considerable solar input is minimized and the energy bill is Rs.17.76/.

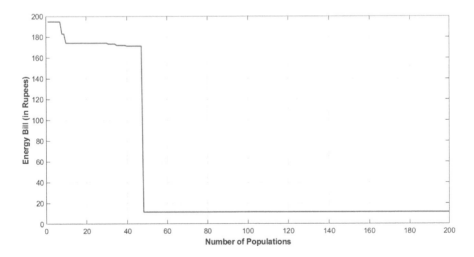

FIGURE 9.2 Economic energy bill calculations over the number of populations (weekday tariff is considered).

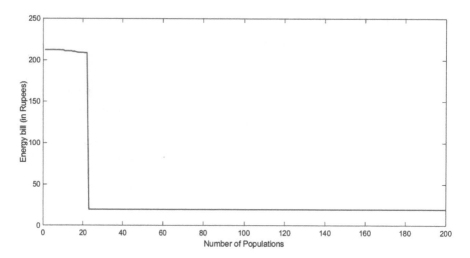

FIGURE 9.3   Economic energy bill calculations over the number of populations (weekday tariff is considered).

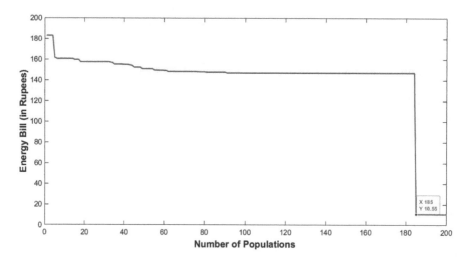

FIGURE 9.4   Economic energy bill calculations over the number of populations (weekday tariff with solar input is also considered).

Case Study D: Minimize the energy bill based on the weekend tariff with considerable solar input and the refrigerator is fixed in all slots. The energy bill of the scheduled appliances over the number of iterations is shown in Figure 9.5. Hour-wise, the appliances are scheduled in the given problem. Among the 24-hour slot, the first slot appliances are 2, 4, 5, 7, 8, and 9 switched on that time, which is shown in Figure 9.6. The analysis can be extended according to the customer needs.

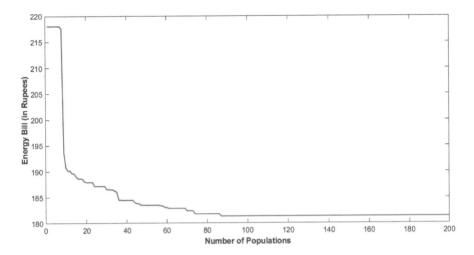

FIGURE 9.5 Economic energy bill calculations over the number of populations (weekend tariff with solar input is also considered and refrigerator is fixed all hours).

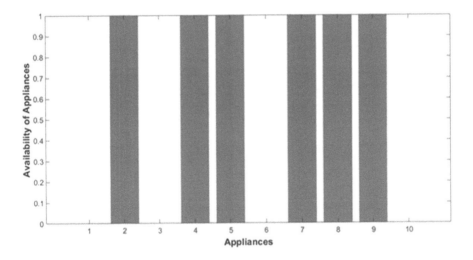

FIGURE 9.6 The switched-on appliances at the first hour by considering weekend tariff with solar input and refrigerator is fixed all hours (12 midnight).

9.5 CONCLUSION

In this work, the home energy tariff is minimized according to the conditions specified. The tailor-made problem could be framed based on the customer orientation. Hence, multiple and simple case studies are projected. The results are proofed that the appliances could be scheduled according to their operations.

REFERENCES

[1] A Schmidt and K Van Laeroven, "How to build smart appliances", *IEEE Personal Communications* 8, 4, 2001, 66–71.

[2] AC Batista and LS Batisa, "Demand side management using a multi-criteria ε-constraint based exact approach", *Expert Systems with Applications* 99, 1, 2018, 180–192.

[3] M Beaudin and H Zareipour, "Home energy management systems: a review of modelling and complexity", *Renewable and Sustainable Energy Review* 45, 2015, 318–335.

[4] FA Qayyum, M Naeem, AS Khwaja, A Anpalagan, L Guan, and B Venkatesh, "Appliance scheduling optimization in smart home networks", *IEEE Access* 3, 2015, 2176–2190.

10 IoT-Based Smart Metering

Parth Bhargav, Fahad Nishat, Umar Ahmad Ansari, and O.V. Gnana Swathika
Vellore Institute of Technology

CONTENTS

10.1 INTRODUCTION

The concept of Internet of Things (IoT) allows people to connect basic devices that we use in our daily life to link with each other via the Internet. Devices connected and governed through the IoT protocols can be remotely administered and analyzed. IoT conception provides fundamental infrastructure and possibilities to establish a bridge between the real-world and computing systems [1]. With the rapid growth of more and more wireless devices on the market, this concept becomes more and more important. It links hardware peripheral to one another via Internet. The Manufacturer Espressif Systems 8266 Wireless module used within the system provides a connection to the Internet in the system [2] . Today, the demand for electricity of the whole population is increasing at a constant rate guide for various domains like agriculture, industry, home use/domestic use, hospitals, etc. The entire population's demand for electricity is growing at a constant rate, guided by different purposes like agriculture, industry, household use, hospitals, etc. As a result, handling power maintenance and demand becomes more and more complex; also, there is a pressing need to save as much power as possible [3]. As the ordinant dictation for electricity from the new generation of inhabitants continues to increase, corresponding technological improvements are required. The proposed system provides a technical improvement for ordinary electricity meters using IoT technology [4]. In addition, we must also solve other problems, such as power larceny and meter tampering, which in turn causes profitability losses for the country [5]. Monitoring and optimizing power consumption and minimizing power waste are the main goals for achieving a better power structure and network. For billing, an individual human is required for accessing each customer's electricity meter and engenders a bill by taking unit readings from the electricity meter [6–9]. In order to eradicate this time-consuming process and solve all the given restrictions, we have created a system from scratch contingent on IoT technology. Design of smart meters based on Wi-Fi systems is based on the three main goals:

- Provide instant automatic meter reading function basis.
- Use electricity in the best way.
- Cut down the power wastage.

Correspondingly, the system should additionally be a subsidiary of the server. Consequently, the concrete system can fundamentally be relegated into two ways predicated on the server.

- End user/back-end.
- Front end.

The data from the back-end system is exhibited on a web page, accessible to both consumers and accommodation providers. This system is depicted on the Arduino microcontroller [10]. It can be divided into three components in structure: controller, larceny detection circuit, and wireless fidelity unit. The controller executes rudimentary calculations and acts upon the information. Larceny detection circuit can provide details about any meter interference, and the wireless fidelity unit will play the most paramount role to send information from the controller via Internet. The code required for the Arduino controller is Programming on the Arduino software IDE and is a prerequisite for running on the Arduino board [11–14]. The wireless fidelity(Wi-Fi) module is the important and chief component utilized in the complete IoT operations. The core of the Arduino board is to provide connections between the different components of the depicted system. The Arduino UNO board is predicated on the AT mega 328p processor [15]. This is the core of the system and is obligatory for rudimentary operations that must be performed (for example, automatic electricity bills and tamper detection input for tampering circuits) [16]. The load represents the equipment that needs to be energized. The AC supply of power is connected to the system through a transformer for supplying a power to the system [17]. Instrument also connects to the system for automatic use family [18]. Then, read the reading from the meter Process and update via wireless fidelity via Espressif Systems 8266 Wireless Fidelity module.

In case any tampering is encountered by the system, it will update the following condition: it will transmit the packets over the cloud on the required web page and web page is used to exhibit energy readings. In every 5 seconds, it sends the pulse from the energy meter to the cloud for processing, and it also updates energy reading timeline on the web page in every 5 seconds. The energy reading is exhibited on the LCD (liquid crystal) exhibit [19]. In case any interference has been done in the energy meter, the buzzer will make a loud sound noise.

All information from the system is facile. It can be found on a web page called Thingspeak.com [13].

The advantages are as follows:

- Reduce energy waste.
- Prevent power shortages during the dry season.
- Make each customer a protector of their own interests power supply.
- Provide authentic-time bill monitoring.
- Reduce receiving timeline.

Application field

- Residential and commercial buildings in public energy supply system
- Government energy plant

10.2 PROPOSED SYSTEM

Among the suggested techniques, customers can handle and obtain the use of vitality by understanding the use of its vitality frequently or once a while [20]. This scheme provides many paths, including the interchange linking the two utilities, and buyers additionally provide different customers' facility to ignore payment power and vibrant supply will be cut off. In terms of utilities, when the user will pay the bill, the front-end user supply also reconnects. The suggested system has a feature where the perpetual alert messages are sent to the consumer which includes payment details and approved payments [21]. To minimize the energy consumption, limits are set carefully while using appliances and devices. If the electronic watch is faulty, a notification will additionally be sent to the utilizer or user [22]. IoT server Thingspeak. com is utilized as a cloud server. Thingspeak.com is the first online builder/implement for engendering IoT projects [12]. The voltage and current values are perpetually stored in the server. Alarms can be scheduled on the server. Most of the proposed framework will play a role in two aspect modes.

1. Unmanned mode
2. Physical mode

Preference of Choices
Unmanned mode: In this one, it crosses the inhibition. The contrivance will automatically switch off. Contrivance culled through utilizer accommodation.

Physical mode: In this mode, manually turn on the switch manual mode. In manual mode, the customer can consume as much presence as possible well-kenned customers.

Problem scrutiny: Power panel is being used in a physical procedure, despite the facts that there are many worries combined with it [15]. Given human error occurred after charging personnel fees, this is customer problem to obtain adjustment from vitality supply board. Consider everything that the customer needs to do. In the workplace, keep in line and get redressed. The issue is the result of human intercession. Keep strategy. Keep away from human error construction procedures. In this incipient era, frame came into being [23–25].

10.3 OUTCOME

Implementation: Keenly intellective meter utilizes Wi-Fi. The module can be facilely divided into two components. The first is the physical part, while the second part is the physical part web page.

Physical part: It is composed of Arduino board and Espressif Systems 8266 Wireless Fidelity module, LCD exhibit, alarm, and power transmission. Arduino Uno enhancement board Arduino is a microcontroller board, which is predicated on AT mega 328P. It can be powered by the puissance supply at the potency outlet. You can utilize an AC-to-DC adapter or with a battery [7].

Wireless module: Espressif Systems manufactures wireless fidelity module low-cost components made by manufacturers. Microcontroller module is capable of performing wireless networking. In Espressif Systems 8266, the Wi-Fi module is a system on a chip with the following functions: 2.4 GHz range. It utilizes a 32-bit RISC CPU running at 80. It is predicated on TCP/IP (Transmission Control Protocol). It is the most consequential component in the system to perform IoT operations. It has 64-KB boot ROM, 64-KB Injunctive sanction RAM, and 96-KB data RAM. Wireless Fidelity unit of the project is used for execution of IoT operation by sending electricity meter data to a web page.

LCD exhibit: LCD screen is an electronic exhibit module, found in a lot of applications. 16×2 denotes that 16 can be exhibited in each line of characters, and there are two such lines. In this LCD, each character is exhibited in a matrix of 5×7 pixels. 11, 12, 13, and 14 pins of the exhibit are utilized as data pins of the Arduino interface, which is used to exhibit wattage.

Working of E-Meter: The vitality and purport of the evaluation circuit board are kenned as a vitality meter. Vitality is aggregation [2]. The puissance is consumed and utilized by the heap in a categorical duration. It is utilized as a component of housing and machinery, Alternating current-driven circuits are used to estimate power utilization. The meter is more cheap and gives more delicate reading. The fundamental unit of energy is watts. One kilowatt is 1 kW. When utilizing 1 kW at $60°$ minutes, it is considered a unit that devours vitality. These meters measure instantaneous voltage and flow. This one integrates forces in a period of vitality utilized in that era and age.

10.4 HARDWARE-USED INTERPRETATION

Arduino AT-MEGA: It is an AVR-compliant low-powered 8-bit CMOS microcontroller. It has upgraded minimized injunctive authorization set computer (RISC) design. It has less power consumption and minute size [26].

EEPROM: EEPROM recollection withal has a dedicated chip Expunge mode; you can expunge the entire chip within 10 ms. Compared to Expunge EPROM, you can expunge and reprogram The contrivance is in the circuit. However, EEPROM is at most sumptuous and least dense ROM [27].

Regulated power supply: Power plan includes rugged power transformer (additionally can Detachment between information and its benefits) and dissipate the layout of the controller circuit.

Controller: The circuit can contain a single Zener diode or three diodes. The terminal directly arranges the controller to distribute the required resultant voltage [12].

Liquid crystal exhibit: Supplementally, it has 64 bytes of character engenderer (CG) RAM. This recollection is characters that are utilized for customer characteristics [20].

RTC: Genuine Time Clock (RTC) is utilized for data resetting and store in online and offline modes. The most diminutive involute shape includes a loop utilized as an open circuit for the solenoid, and close the switch contact [28].

The Internet of Things server: Thingspeak.com is utilized as a cloud service provider. Thingspeak is the first online cloud service provider implemented for engendering IoT projects. The current and voltage values are perpetually stored in the server. Alarms can also be programmed on the server [29].

Wireless fidelity module: Espressif Systems 8266 is a Wireless Fidelity Module Felicitous for integrating wireless fidelity usability to the current version, Take a microcontroller adventure through GM Asynchronous transmitter-receiver (UART) serial sodality. The module can even be reinvented to solve as an unrestrained contrivance cognate to wireless fidelity [30].

Voltage electric eye: DC the state of being kinetic from the AC frame In order to contribute to the microcontroller, we are capitalizing on this voltage detection circuit. The circuit gives a precise technology for making this DC logo.

Detection voltage: By utilizing a voltage transformer, the denotement obtained is modified in the main op-amp stage, and the second operational amplifier is arranged [18].

Current sensor: By detecting the current, current transformer rectifies the main operation amplifier stage and booster for the second operational amplifier arrangement [31].

10.5 CONCLUSION

The reason for the design of electricity meters predicated on the IoT is to minimize indoor power consumption. It eschews human intervention, truncates costs, and preserves manpower. Both automatic and manual are available. The computerization can minimize the total work cost and make the framework more efficacious and precise [32]. This system is mainly utilized in astute cities with public places wireless fidelity hotspot area. The project is predicated in the cyber world. The concept of things is designed to supersede old energy instruments with advanced implementation. It can be used in automatic power reading, through which its power can be optimized, thereby truncating power waste. Upload the meter reading to Thingspeak. com channels for concrete energy utilization. The server can view the meter customer. In conclusion, in the era of astute city development, this project fixates on connection and networking factors of the IoT. In this project, energy consumption, the calculation predicated on the calibration pulse count is the use of PIC16F *&A MCU design and implementation of the embedded system domain. In the proposed paper, IoT and PLC-predicated meter reading system aim that the energy electricity meter readings are perpetually monitored. As long as the customer does not pay the monthly bill, the accommodation provider can disconnect the potency supply, eliminating manual intervention, providing efficacious meter readings, and averting billing errors. The project has achieved the following goals: (i) it is facile to access consumer information from electricity meters through the IoT; (ii) genuine-time larceny detection is on the utilizer side; (iii) LCD exhibits energy squander unit and temperature; (iv) disassociates the accommodation from the remote server. Future enhanced functions in this system, IoT electricity meter, use the wireless fidelity module to access consumption; it will avail consumers evade nonessential utilization of electricity. The system performance can be enhanced in the following ways. Connect all home appliances to the IoT. Consequently, the following goals can be achieved in the future to preserve costs power and eschew larceny:

1. We can build an IoT system for users to monitor energy consumption and pay bills online.
2. We can build a system where users can enable short messaging service in their mobile phone when they want or when he/she exceeds the doorway of electricity consumption.

REFERENCES

[1] Ashna K & George SN, 2013. GSM based automatic energy meter reading system with instant billing, *Proceeding of International Multi Conference on Automation, Computing, Communication, Control and Compressed Sensing (Imac4S)*, 65(72), 22–23.

[2] Sasanenikita N, 2017. IOT based energy meter billing and monitoring system, *International Research Journal of Advanced Engineering and Science*.

[3] Swathika, OG, Kanimozhi, G, Umamaheswari, E, Rujay, S and Saha, S, 2021. IoT-Based energy management system with data logging capability. In *Advances in Smart Grid Technology, Springer*, Singapore, pp. 547–555.

[4] Malik NS, Kupzog F, & Sonntag M, 2010. An approach to secure mobile agents in automatic meter reading, *IEEE International Conference on Cyber Worlds, Computer Society* 187–193.

[5] Depuru SSSR, Wang L, & Devabhaktuni V, 2011. Electricity theft: overview, issues, prevention and a smart meter based approach to control theft, *Energy Policy* 39(2), 1007–1015.

[6] Hiwale AP, Gaikwad DS, Dongare AA, & Mhatre PC, 2018. Iot based smart energy monitoring. *International Research Journal of Engineering and Technology (IRJET)* 5(03).

[7] Dong S, Duan S, Yang Q, Zhang J, Li G, & Tao R, 2017. MEMS-based smart gas metering for internet of things. *IEEE Internet of Things Journal* 4(5), 1296–1303.

[8] Singh A & Gupta R, 2018. IoT based smart energy meter. *International Journal of Advance Research and Development* 3(3), 328–331.

[9] Odiyur Vathanam GS, Kalyanasundaram K, Elavarasan RM, Hussain Khahro S, Subramaniam U, Pugazhendhi R, Ramesh M, & Gopalakrishnan RM. A review on effective use of daylight harvesting using intelligent lighting control systems for sustainable office buildings in India, 2021. *Sustainability* 13(9), 4973.

[10] Pandit S, 2017. Smart energy meter using internet of things (IOT), *VJER Vishwakarma Journal of Engineering Research* 1(2).

[11] Nyirendre CN, Nyandowe I, & Shitumbapo L, 2016. "A comparison of the collection tree protocol (CTP) and AODV routing protocol for a smart water metering", pp. 1–8.

[12] Kurde A & Kulkarni V, 2016. IOT based smart power metering. *International Journal of Scientific and Research Publications* 6(9), 411–415.

[13] Bhilare R & Mali S, 2015. IoT based smart home with real time E-metering using E-controller. In *2015 Annual IEEE India Conference (INDICON)*, IEEE, pp. 1–6.

[14] Yaghmaee MH, & Hejazi H, 2018. Design and implementation of an internet of things based smart energy metering. In *2018 IEEE International Conference on Smart Energy Grid Engineering (SEGE)*, IEEE, pp. 191–194.

[15] Maitra S, 2008. Embedded energy meter-a new concept to measure the energy consumed by a consumer and to pay the bill, In *Power System Technology and IEEE Power India Conference*, IEEE, pp. 1–8.

[16] Yaacoub E & Abu-Dayya A, 2014. Automatic meter reading in the smart grid using contention based random access over the free cellular spectrum, *Computer Networks*.

[17] Sehgal VK, Panda N, Handa NR, Naval S, & Goel V, 2010. Electronic energy meter with instant billing, *Fourth UKSim European Symposium on Computer Modeling and Simulation (EMS)* 27–31.

[18] Arasteh H, Hosseinnezhad V, Loia V, Tommasetti A, Troisi O, Shafie Khan M, & Siano P, 2016. "IoT based smart cities: a survey", IEEE.

[19] Pang C, Vyatkin V, Deng Y, & Sorouri M, 2013. "Virtual smart metering in automation and simulation of energy efficient lightning system", IEEE.

[20] Ainabekova G, Bayanbayeva Z, Joldasbekova B, & Zhaksylykov A, 2018. The author in esthetic activity and the functional text (on the basis of V. Mikhaylov's narrative "The chronicle of the great jute"), *Opción, Año* 33, 63–80.

[21] Shrouf F, Ordieres J, & Miragliotta G, 2014. Smart factories in Industry 4.0: a review of the concept and of energy management approached in production based on the internet of things paradigm. In *2014 IEEE international conference on industrial engineering and engineering management*, IEEE, pp. 697–701.

[22] Ahmed E, Yaqoob I, Gani A, Imran M, & Guizani M, 2016. Internet-of-things-based smart environments: state of the art, taxonomy, and open research challenges. *IEEE Wireless Communications* 23(5), 10–16.

[23] Rastogi S, Sharma M, & Varshney P, 2016. Internet of things based smart electricity meters. *International Journal of Computer Applications* 133(8), 13–16.

[24] Avancini DB, Rodrigues JJ, Rabêlo RA, Das AK, Kozlov S, & Solic P, 2021. A new IoT-based smart energy meter for smart grids. *International Journal of Energy Research* 45(1), 189–202.

[25] Joshi DSA, Kolvekar S, Raj YR, & Singh SS, 2016. IoT based smart energy meter. *Bonfring International Journal of Research in Communication Engineering* 6, 89–91.

[26] Prasad SG, 2017. IOT based energy meter, *International Journal of Recent Trends in Engineering & Research (IJRTER)*.

[27] Praveen MP, 2011. KSEB to introduce SMS-based fault maintenance system, *The Hindu News*.

[28] Bhimte A, Mathew RK, & Kumaravel S, 2015. "Development of smart energy meter in labview for power distribution systems", In *2015 Annual IEEE India Conference (INDICON)*, IEEE, pp. 1–6.

[29] Barman BK, Yadav SN, Kumar S, & Gope S, 2018. IOT based smart energy meter for efficient energy utilization in smart grid. In *2018 2nd International Conference on Power, Energy and Environment: Towards Smart Technology (ICEPE)*, IEEE, pp. 1–5.

[30] Mir SH, Ashruf S, Bhat Y, & Beigh N, 2019. Review on smart electric metering system based on GSM/IOT. *Asian Journal of Electrical Sciences* 8(1), 1–6.

[31] Yesembayeva Z, 2018. Determination of the pedagogical conditions for forming the readiness of future primary school teachers, *Opción, Año* 33, 475–499.

[32] Landis JW, 2014. "Billing metering using sampled values according lEe 61850-9-2 for substations", IEEE.

11 Smart Decentralized Control Approach for Energy Management in Smart Homes with EV Load

M. Subashini and V. Sumathi
Vellore Institute of Technology

CONTENTS

ABBREVIATIONS

AC	Alternating Current
BSS	Battery Storage System
CC	Constant Current
CV	Constant Voltage
DC	Direct Current
DOD	Depth of Discharge
EB	Electricity Board
EV	Electric Vehicle
IST	Indian Standard Time

DOI: 10.1201/9781003240853-11

MNRE   Ministry of New and Renewable Energy
MOSFET   Metal Oxide Semiconductor Field-Effect Transistor
MPP   Maximum Power Point
PLL   Phase-Locked Loop
PWM   Pulse Width Modulation
SOC   State of Charge

11.1   INTRODUCTION: IMPORTANCE OF ENERGY SAVING

Energy saving and usage of renewables are two of the major goals in the energy sector to protect our environment, as it indirectly minimizes the usage of fossil fuels. On the other hand, it is predicted that the residential consumption in India will increase around 85% by 2050 as in Ref. [1]. Thus, it will become mandatory to save energy in all possible ways. In order to reduce energy consumption, and to incorporate the green energy resources into our system, a deep analysis is required in the load side. A major energy loss happens especially in uncontrolled and less-maintained equipment and also operating the devices at light load conditions. This paper proposes a smart energy-saving method that can be easily adopted in smart home environments; also, it purports smart control strategies for EV battery charging.

11.1.1   LOAD CATEGORIZATION AND CONTROL STRATEGY

Design process starts with the categorization of the domestic electrical equipment, based on its energy requirements, into three, namely, less-impact loads (DC loads, especially lights and fans), medium-impact loads (refrigerator, washing machine, air conditioner etc.), and high-impact loads (EV battery).

To alleviate the extra burden on the initial investment cost of PV installation, its size is considered to meet only the house demand (excluding EV charging) and to charge the backup battery.

The controller initiates the charging process of EV battery autonomously during the off-peak hours or when the house demand is very low. The parameters that influence the charging decision of the controller are grid conditions, house demand, PV array output, SOC of the backup battery, and the SOC of the EV battery. This intelligent system mitigates the effect of sudden rise of domestic energy demand while adopting EV charging as part of the domestic load. Over a period of time, the initial investment for the renewable energy can be met with reduced running cost. This control algorithm will be more effective in countries like USA, which uses the "Time of day" tariff. Section 11.2 deals with the design of primary resources, PV array size calculation, and backup battery size calculation. Section 11.3 covers the design of inverter/converter required for other subsystems. Controller design is elaborated in Section 11.4. Section 11.5 is dedicated to results and conclusions.

11.2   PROPOSED SYSTEM

The energy demand for a smart home can be met with two resources, namely, renewable source (PV) and utility grid. The priority order of power sources for utilization are PV and then the grid. Designing a net zero energy home is a best solution to

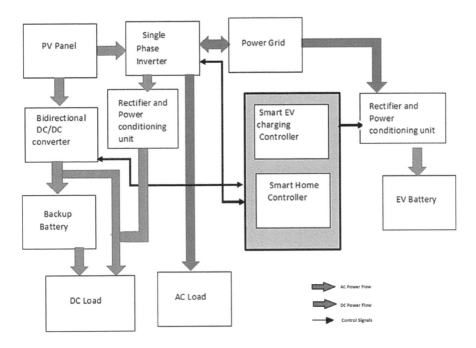

FIGURE 11.1 The block diagram of the entire smart home energy management system.

self-sustainability. But, due to cost constraints, here we have limited our design of PV to supply the house load. The EV load taken from the grid is scheduled during off-peak hours. Figure 11.1 shows the block diagram of the entire Smart Home Energy Management System. The system is divided into three major subsystems, namely, inverter subsystem, backup battery subsystem, and EV battery charging subsystem. The design considerations for the main system resources are discussed below.

11.2.1 SIZING OF PV PANEL

A smart home, which is the integral part of a smart grid, and located in Chennai, was chosen for the analysis, and its monthly consumption is assumed to be around 200 units. To derive the design parameters, TATA Nexon Electrical car [2] and Renesola PV panel model (250 W) [3] were considered for the design calculations.

Chennai is located at 13.0827°N, 80.2707°E, and its minimum irradiation received during a day is 5.5 as per data provided by reference [4]. Design calculations for sizing of PV to meet the house demand are performed using reference [5,6]. Essential parameters for system modeling are given in Table 11.1

$$W_{hpv} = \frac{Wh_{\text{load}} \, / \, \text{month}}{30} \tag{11.1}$$

where

W_{hpv} – the watt hour of the PV panel required

Wh_{load} – the energy requirement of house load in terms of Wh

TABLE 11.1

Name Plate Details of Renesola PV Panel Model (250 W)

Sl. No.	Parameter	Value
1	Maximum power (P_{max})	250 W
2	V_{oc}	37.40 V
3	V_{mp}	30.1 V
4	Maximum system voltage	1,000 V DC
5	Dimension ($L*W*H$)	$1,640 \times 992 \times 10$ mm
6	Power tolerance	6/+5 W
7	Short-circuit current	8.83 A
8	Maximum power current (Imp)	8.32 A
9	Maximum series fuse rating	20 A
10	Weight	19 kg

Let's assume that the efficiency of solar panel is approximately 16%. Then, the intrinsic area (active area in sq. meter to produce the required power) of solar panel is calculated by the following formula:

$$A_{intrinsic} = \frac{W_{hpv}}{\eta * H_{atmin}} \qquad (11.2)$$

where

H_{atmin} – the minimum value of daily solar irradiation received during a day by a squared meter of surface (Wh/m²)

η – the efficiency of solar panel

$A_{intrinsic}$ – the area of solar panel required to generate the required amount of energy in sq. m

A_{estate} – the total area required for installation

After the calculation of intrinsic area, the total area required for installation of solar panels, i.e., the estate area (area required for maintenance, etc., is nearly 1.3 times of that required value)

$$A_{estate} = 1.3 * A_{intrinsic} \qquad (11.3)$$

The dimension of the single module is 1.63 m². The panel count can be calculated with the simple formula:

$$\frac{A_{estate}}{\text{Area of single module}} \qquad (11.4)$$

If the calculations were performed for the monthly consumption of 200 units, approximately six PV panels were required. Energy generated in terms of number of units by these six panels per day can be enumerated as follows:

$$\text{Efficiency}(\eta) = \frac{V_{mp} * I_{mp}}{(1.63 * 1000)} \tag{11.5}$$

$$W_{hpv} = \eta * A_{\text{intrinsic}} * H_{\text{atmin}} \tag{11.6}$$

where
V_m – the voltage at MPP
I_m – the current at MPP

Approximate generation in 1-month duration is 248 units.

11.2.2 Sizing of Backup Battery

This section discusses the calculation of sizing of backup battery for energy storage [6]. PV energy generated during the day time can be utilized to supply the day demands and back up the excess in the battery. During night hours, battery power is used to manage the load. Among the many energy storage systems, battery storage system (BSS) is the most successful technology for small size backups [7].

Design calculations were performed considering lead acid (tubular) [8,9] battery bank for energy storage (specifications are listed in Table 11.2) because they are the stronger counterparts of lead acid batteries with longer storage of energy, greater number of cycle time, and deep cycle discharge [10].

Assuming zero days of autonomy and Ah capacity of the backup battery are just to meet out the night demands, then Watt hour requirement of the backup battery (Wh_{load}) is

$$Wh_{\text{load}} = Wh_{\text{night}} / \eta_B \tag{11.7}$$

Then, the ampere hour of battery is calculated as:

$$Ah_{BB} = Wh_{\text{night}} / (DOD * V_{\text{bat}}) \tag{11.8}$$

TABLE 11.2
Specifications of Lead Acid Tubular Battery

Sl. No.	Parameter	Value
1	Nominal voltage	24
2	Efficiency	90%
3	Depth of discharge	80%
4	No. of charge/discharge cycles	2,000

where
  Ah_{BB} – the ampere hour of the backup battery
  DOD – the depth of discharge of the backup battery
  V_{bat} – the nominal voltage of the backup battery
  Wh_{night} – the night demand of the smart home
  η_B – the efficiency of the backup battery

For a night load of around 1,000 Wh with lead acid tubular battery storage, the Ah_{bat} requirement size is up to 70 Ah, including the added 20% safety marginal to the capacity.

Consider the maximum load current and average load current of $I_{lmax} = 6$ A and $I_{lavg} = 3$ A, respectively, for house load. A 70-Ah battery with C_{20} rating (discharging at the rate of 1/20 value of Ah capacity) is selected (70/20 = 3.5 Amps). The lesser the value of C rate, the more the discharge time.

11.3   DESIGN OF SUBSYSTEMS

This section covers the detailed design procedure for developing the three subsystems, namely, the inverter subsystem, backup battery subsystem, and EV battery charging subsystem.

11.3.1   INVERTER SUBSYSTEM

Inverter subsystem ensures stable grid connectivity and protection for the PV array (situations like islanding). A DC link capacitor connects the PV module to the utility grid though a single-phase inverter. Full-bridge MOSFET inverter circuit is used and made to operate in synchronization with the grid. Here, the PLL-based grid synchronization method is adopted. In voltage-controlled grid synchronization, inverter output voltage and frequency were matched with grid voltage and frequency [11,12]. Figure 11.2 shows the block diagram of inverter subsystem. PWM is generated based on the parameters such as modulation index (m), angular frequency of grid voltage (ω), and the required phase shift (shift$_k$). The formulas were listed below.

$$m = V_{grid} / V_{inv} \tag{11.9}$$

$$x = \omega \text{ in rad/seconds equation} \tag{11.10}$$

$$a = m * \left(\sin(x) + \text{shift}_k \right) \tag{11.11}$$

The calculated value "a" is considered as duty to generate the required PWM for the inverter.

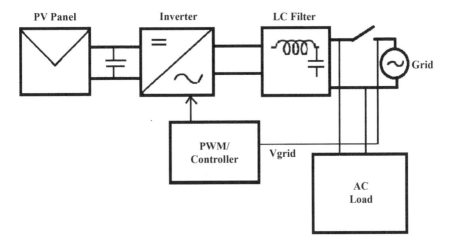

FIGURE 11.2   Block diagram of the inverter subsystem.

11.3.2   FILTER CIRCUIT DESIGN FOR INVERTER

Filter design equations were given below. As per the system requirements, various parameters were calculated using [13] and listed in Table 11.3.

$$f_c < \frac{1}{10} * f_{sw} \tag{11.12}$$

$$L < \frac{0.03 * V_{inv}}{2 * \pi * f * I_{lmax}} \tag{11.13}$$

$$C = \frac{1}{\left(2 * \pi * f_c\right)^2 * L} \tag{11.14}$$

where
  f_c – the carrier frequency in Hz
  f_{sw} – the switching frequency (5 KHz) in Hz
  I_{lmax} – the maximum value of load current in Amps
  f – the frequency of the inverter output voltage in Hz

TABLE 11.3
Parameters of Inverter Filter

Sl. No.	Parameter	Value
1.	V_{inv}	311 V peak value
2.	I_{lmax}	30 A
3.	f_c	<500 Hz
4.	L	1 mH
5.	C	100 µF

11.3.3 Backup Battery Subsystem

The backup battery subsystem consists of two DC/DC converter circuits: (i) buck converter and (ii) two-switch bidirectional DC/DC converter.

The buck converter [14,15] used here is to match the PV array output with the DC bus (48 V) From the DC bus, the DC loads were supplied power when the PV output is ON.

The block diagram of the backup battery subsystem is given in Figure 11.3.

The bidirectional two-switch buck-boost DC/DC converter is used between the DC load and the backup battery. This bidirectional converter is a modified version of the common buck-boost converter. The design equations can be used with a slight modification [16,17]. The parameters of this bidirectional converter are calculated and listed in Table 11.4.

Control algorithm (explained in Section 11.4) automatically starts or stops the backup battery charging and discharging cycles.

There are two stages in the backup battery charging action: one with constant current charging (CC) up to 80% of SOC and then constant voltage control (CV) [18] above 80% of SOC. The control pulses for the bidirectional converter are generated by using a double-loop PI controller.

TABLE 11.4
Parameters of Bidirectional Converter

Sl. No.	Parameter	Value
1.	Nominal voltage of the battery	24 V
2.	Load voltage	48 V
3.	Load current	5 A
4.	Switching frequency (f_{sw})	5,000 Hz
5.	L	0.2 mH
6.	C	200 μF

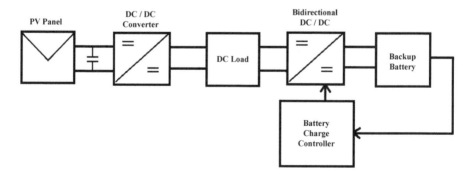

FIGURE 11.3 Block diagram of the backup battery charging/discharging control subsystem.

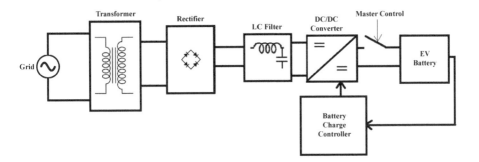

FIGURE 11.4   EV battery charging subsystem.

11.3.3.1 EV Battery Charging Subsystem

The EV battery is charged by the battery charging subsystem custom-designed charging circuit for improved performance and better control. The system is enriched with required safety protocols. The charging circuit draws power from the utility grid and conditions the same and supplies the stable power for charging the EV battery. The power conditioning unit includes step-up transformer 230–340 V, rectifier, and filter unit along with a boost converter [19, 20]. DC charging unit provides 320 V (nominal DC voltage to EV battery) and provides 15 A pulse current for charging the battery. Block diagram of EV battery charging subsystem is shown in Figure 11.4.

Master control switch is controlled by the smart control algorithm, which is discussed in the next section.

From the data sheet of Tata Nexon EV specifications, its on-board charger can handle a maximum power of 3.3 kW only. Domestic installations can go up to 7 kW with level 2 charging. Here, extra hardware-controlled charging can be carried out with pulse current ranging from 15 to 21 A. Parameters such as charge current, charging time, and charging duration can be controlled. In order to improve the efficiency of charging, from zero to 80% SOC_EV, the battery is charged at constant current control mode and from 80% to 100% in the constant voltage mode.

11.4 MASTER CONTROLLER

Master controller is the brain of the system and works autonomously and takes decisions based on the various sensed parameters listed below:

1. PV output (PV_OUT)
2. Backup battery SOC (SOC_BB)
3. EV battery SOC (SOC_EV)

Based on the above-listed parameter values, the controller takes decisions and operates switches to open or close in order to accomplish the task. Here, the PV_OUT level is divided into fair-to-middling, ample, and plentiful [21]. Switches S1 to S5 are used, and their functions are listed below:

Inverter control switch (S1): When the PV_OUT is plentiful, this switch is closed to power the AC loads. Otherwise, the grid manages the AC loads.

DC load power-up switch (S2): When the PV_OUT is more or less equal to fair-to-middling, the switch is closed and powers direct current devices.

Backup battery control switch (S3): The decision to open or close this switch is based on two parameters, PV_OUT and SOC_BB. If PV_OUT reaches the ample, and also if the SOC_BB < 20%, this switch closes and charges

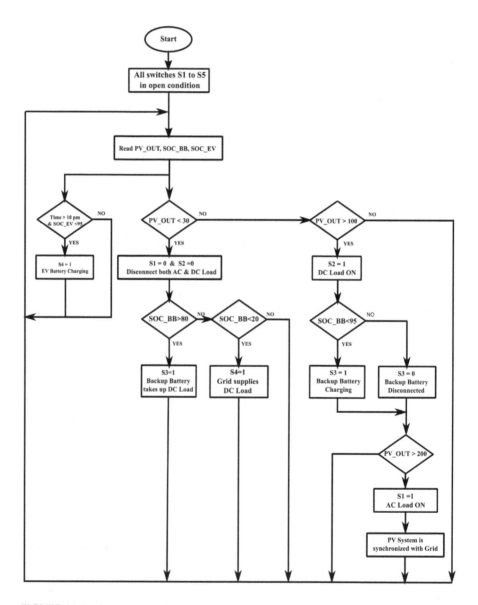

FIGURE 11.5 System control flowchart.

the backup battery. Or else, if SOC > 95%, then the battery is disconnected during day. During night hours and low generation times, the backup battery takes the DC load.

As worst-case scenario, if both PV and backup battery were down (for the conditions S2 and S3 are in open), S4 is closed and the grid will supply power to the DC battery.

Master control switch (S5): It closes when the SOC_EV < 95% and the IST is between 10 p.m. IST and 5 a.m. IST (during night hours). The flowchart of the intelligent control algorithm is presented in Figure 11.5.

Control algorithm is explained in the following simple steps:

1. Automatically switch the AC load between the grid and PV source based on the PV output.
2. DC load is also automatically shuttled back and forth between PV source and backup battery.
3. Also, charging decision of EV battery and backup battery is based on their respective SOCs.

11.5   RESULTS AND CONCLUSION

The entire system is simulated in MATLAB Simulink, and the parameters of interest were showcased in the wave forms below. Figures 11.6–11.8 portray the output of inverter, backup battery, and EV charging wave forms, respectively. The EV battery charging unit is designed with variable charging current between boundaries 15 and 22 A during CC charging mode; thereby, the charging duration can be varied to some extent. From the data sheet of Tata Nexon, the range of the vehicle is 312 km for a full charge and requires around 30.2 units.

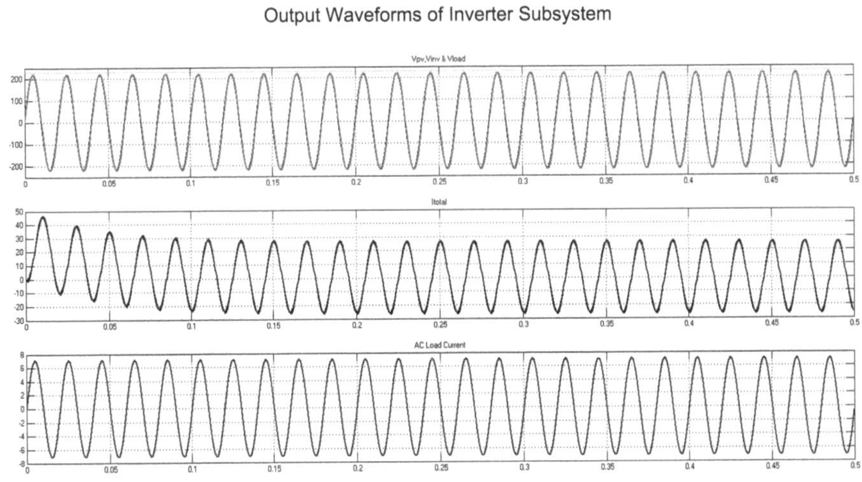

FIGURE 11.6   Inverter subsystem waveforms.

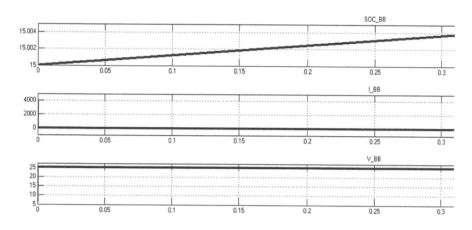

FIGURE 11.7   Backup battery subsystem waveforms.

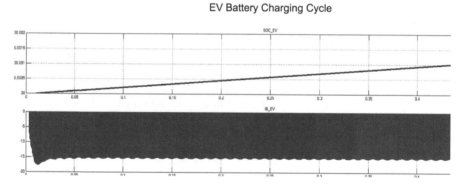

FIGURE 11.8   EV battery charging waveforms.

TABLE 11.5

Approximate Coast of PV Installation Considering the State and MNRE Subsidy

PV	Approx. Cost in Rs	30% MNRE Subsidy in Rs	State Govt. Subsidy in Rs	Net Cost after Subsidy Approx. in Rs
Total cost for 1.5-kW PV panel	90,000	27,000	20,000	43,000

Approximate cost-saving calculations were performed and tabulated [22]. A good saving in monthly consumption and within a few years of use (around 5–6 years), the initial investment can be met if the subsidy for renewable installation is provided. Table 11.5 shows the state and central subsidies for PV installations.

TABLE 11.6
Approximate Calculation of Annual Saving in Rupees

EV Runs an Avg. of km/day	EV Consumption for Two Months in Units	House Consumption for Two Months in Units	Total Consumption for Two Months in Units	EB Bill for Two Months without PV in Rs	Annual Cost without PV in Rs	Annual Revenue Generated from PV in Rs
50	302	400	702	1,912	11,472	8,640

With the calculated value of PV generation (248 units) and house load of 200 units (both assumed to be constant), the annual saving can be calculated. The result shows that the energy saving per annum is in the range of around Rs. 9,000. Table 11.6 shows the calculation of the annual cost saving with PV support.

With intelligent design of home energy management system and some initial investment for renewables, we can make the system energy efficient and cost-effective. The system can be improved with Fuzzy-Neuro controller, solar energy prediction, and load scheduling.

REFERENCES

[1] Plugging in: Electricity Consumption in Indian Homes Policy Research India 31 October 2017.

[2] TATA Motors Nexon EV Brochure, https://nexonev.tatamotors.com.

[3] ReneSola 250 V Panel Rating, https://www.energysage.com/solar-panels/renesola-ltd/247/jc250m-24bb/.

[4] TANGEDCO Solar Irradiance Data, https://www.tangedco.gov.in/linkpdf/solar%20irradiance%20data%20in%20Tamil%20Nadu.pdf.

[5] B. L. Gupta, M. Bhatnagar, J. Mathur, "Optimum sizing of PV panel, battery capacity and insulation thickness for a photovoltaic operated domestic refrigerator", *Sustainable Energy Elsevier Technol. Assess.* 7, 2014, 55–67.

[6] Design of Photovoltaic System NPTEL Course by Prof. L. Umanand Indian Institute of Science, Bangalore.

[7] N.-K. C. Nair, N. Garimella, "Battery energy storage systems: assessment for small-scale renewable energy integration", *Energy Build.* 42, 11, 2010.

[8] R. M. Dell, D. A. J. Rand, *Understanding Batteries*, Royal Society of Chemistry, Cambridge, 2002.

[9] G. J. May, A. Davidson, B. Monahov, "Lead batteries for utility energy storage: a review", *J. Energy Storage* 15, 2018.

[10] E. Banguero, A. Correcher, Á. Pérez-Navarro, F. Morant, A. Aristizabal, "A review on battery charging and discharging control strategies: application to renewable energy systems", *Phys.: Conf. Ser.* 1007, 2018, 012054, doi: 10.1088/1742–6596/1007/1/012054.

[11] P. Goli, W. Shireen, "PV integrated smart charging of PHEVs based on DC link voltage sensing", *IEEE Trans. Smart Grid* 5, 3, 2014.

[12] A. A. Bakar, M. A. N. Amran, S. Salimin, M. K. M. Jamri, A. F. H. A. Gani, "Modeling of single-phase grid-connected using MATLAB/simulink software", *2019 IEEE Student Conference on Research and Development (SCOReD)*, Seri Iskandar, Perak, Malaysia, 2019.

[13] J. Sedo, S. Kascak, "Control of single-phase grid connected inverter system", IEEE, 2016.

[14] "Basic calculation of a buck converter's power stage", Texas Instruments Application Report, 2015.

[15] D. Zhan, Design Considerations for a Bidirectional DC/DC Converter, Reneses White Paper 2018.

[16] X. Xu, C. Zheng, C. Hu, Y. Lu, Q. Wang, "Design of Bi-directional DC/DC converter", *2016 IEEE 11th Conference on Industrial Electronics and Applications (ICIEA)*, Hefei, 2016.

[17] D. Zhan, "Design considerations for a bidirectional DC/DC converter", Reneses White Paper 2018.

[18] M. Tampubolon, L. Pamungkas, Y. C. Hsieh, H. J. Chiu, "Constant voltage and constant current control implementation for electric vehicles (EVs) wireless charger", *IOP Conf. Ser.: J.*

[19] S. Jalbrzykowski, T. Citko, "A bidirectional DC-DC converter for renewable energy systems", *Bull. Polish Acad. Sci. Tech. Sci.* 57, 4, 2009.

[20] M. H. Rashid, *Power Electronics: Circuits, Devices, and Applications*, Pearson, 2009, India.

[21] N. A. Gounden, S. A. Peter, H. Nallandula, S. Krithiga, "Fuzzy logic controller with MPPT using line-commutated inverter for three-phase grid-connected photovoltaic systems", *Renewable Energy* 34, 3, 2009.

[22] TANGEDCO Tariff Calculator, Web address https://docs.zoho.com/sheet/published. do?rid=s0uyx795c9cec62134abc91dd5a1a144993a9.

12 Design and Implementation of Prototype for Smart Home Using Internet of Things and Cloud

V. Vijeya Kaveri
Sri Krishna College of Engineering and Technology

V. Meenakshi
Sathyabama Institute of Science and Technology

CONTENTS

12.1 INTRODUCTION

Due to the comfort provided, the 21st-century homes are much self-controlled and automated. The users are able to monitor the variety of electrical appliances with the help of the automation system. The installation is very well-planned during the construction of the building as the home automation system is centered on wired communication. Conversely, wireless systems are often of incredible assistance for automation, which has made the implementation of home automation very easy.

12.2 BENEFITS OF HOME AUTOMATION SYSTEM

In the technically advanced era, the usage of the wireless systems like Wi-fi has become more common. Several benefits are acquired by the use of wireless technologies, which is impossible in a wired network done at home and building automation system as well.

12.2.1 COST OF INSTALLATION

Compared to wireless systems [1–3], installation costs are high in wired systems, as cabling must be performed, including the expense of the fabric and therefore the skilled laying of cables (e.g., into walls).

12.2.2 SCALABILITY AND EXTENSION OF THE SYSTEM

Wireless network implementation is more desirable if there is a requirement for network expansion or any change in requirements [4–6]. Wireless network is opposite to wired installations in which extension of cabling is extremely difficult. This enables wireless network installations a seminal investment.

12.2.3 MOBILE PHONE INTEGRATION

With the assistance of wireless networks, integration of mobile devices like PDAs and smartphones with the house automation system becomes probable all over the place and at any time based on the reachability of the device. The wireless technology is suitable for new installations [7–9], in spite of being suitable for renovation and refurbishment.

12.3 RELATED WORKS

The authors have suggested in paper [10–13] a home automation system, which has been wont to remotely control the lights and other home appliances inside their home with the support of touch mobile devices, cloud networking, wireless connectivity, and power cable communication. The proposed system consists of a smartphone application, a remote wireless system also as a PC-based software that gives the client an interface.

The key objective of this paper [14–17] is to plan and apply a controlling and monitoring system for a smart home. In this study, for the main controlling system, LabVIEW software is used to control all the systems within the smart home. It is possible to have a perfect control and is likely to monitor the household appliances from being at anyplace.

The objective of this paper [18–24] is to support senior citizens or physically challenged people. This paper gives the essential idea, to regulate and secure various home appliances using Android phone/tab. The system enables the user to speak with the appliance and, by sending the control sign to the Arduino ADK, controls the devices/sensors.

In this paper [25–30], the authors suggested a newly designed smart home that includes the usage of new technologies such as wireless sensor network and biometric. In the proposed system, biometric devices have been employed at the home entrance for providing more security and authentication during the entering process.

12.4 SYSTEM ANALYSIS

12.4.1 PROBLEM STATEMENT

Cost, flexibility, lack of security, and poor manageability are the major challenges faced by all current home automation systems. This paper's research idea is to develop and introduce an Internet of Things (IoT) and cloud-based home automation system that offers the facility to solve the above challenges faced in the current system. In this framework, IoT is used in the design of a manageable Web interface to monitor all automated home appliances. The proposed framework addresses the flexibility of interconnecting distributed sensors with home automation servers via Wi-fi technology. This allows the option to update or reconfigure the system, leading to a reduction in the cost of implementation.

12.4.2 FEATURE OF THE PROPOSED SYSTEM

The proposed distributed home automation system comprises servers and sensors. The server is in charge of monitoring the different sensors and tracking them. It is easily configured to handle more hardware interface modules (sensors). During this system, Intel Galileo Development Board functions as an Internet server with the support of built-in Wi-fi card port. Home automation system is often accessed from local Web browser using the IP address of the server, or remotely from any PC or PC that is connected on an equivalent LAN or any mobile device linked via the Web to the specified browser IP address to server. The Wi-fi technology is preferred as the network infrastructure that connects the sensors and the servers. It enhances the security of the system and increases the mobility and scalability of the system.

12.5 DESIGN AND IMPLEMENTATION

12.5.1 PROPOSED HOME AUTOMATION SYSTEM

Figure 12.1 depicts the proposed system design model, which contains numerous sensors, such as temperature, gas, LDR, and motion. The Web server (Intel Galileo) is initially connected via Wi-fi to the Internet; once the connection has been established, the parameter reading from various sensors like pa1, pa2, and pa3 has been noted. The threshold limits for the acceptable sensors are set as th1, th2, th3, then

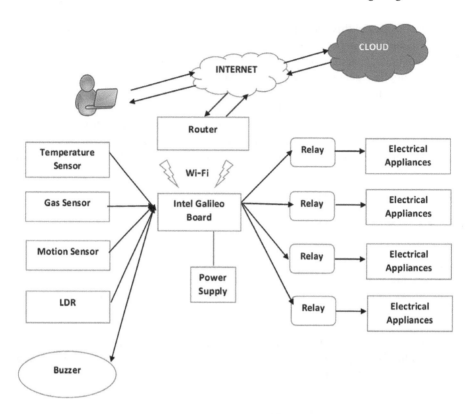

FIGURE 12.1   Proposed home automation system model.

on. The Web server receives the collected data from the sensors and stores them in the cloud, and they are often accessed and analyzed from anywhere all 24×7. If the data obtained from the sensor are found beyond the threshold limit, an alert will be generated by the respective alarm a1, a2, a3, etc., and on the basis of the alarm, the parameters will be controlled.

In the proposed system, the sensors are used to monitor the temperature, leakage of gas, and detection of motion. The data received from the temperature and motion sensor are stored in the cloud, which can be used for analysis. The cooler automatically turns on, if the temperature is found beyond the threshold level, and when the temperature comes to normal level, it will automatically turn off. Similarly, when a gas leakage is detected in the house, the warning sound is automatically raised. The required lights are automatically switched on/off by the detection of the alarm. Using the Web server, users can track all electrical equipment in the home through the Internet. It also offers the possibility of remotely controlling all electrical appliances such as fans, lights, and AC through the Web server's IP address. The functions of the proposed system are to control and monitor the temperature and humidity inside the home, to detect if there is any movement of third person outside the home, to detect fire and smoke due to gas/electrical leakage, and to control the home appliances automatically like fan, light, AC, and heater.

12.5.2 SYSTEM DESIGN

In the proposed system, the user interface (front end) has been designed with the support of HTML with the markup "tags." For storing the data, processing, and managing over the Internet, we have used cloud computing. A traditional utility model is followed by IaaS (or utility processing). Consumers pay for the servers and storage space they use. PaaS helps users, including Google's Application Engine, to create applications within the given framework. SaaS enables the user, through a browser, to use an application on demand. We have used Gmail in our proposed framework to store data that can be accessed from any device with Internet support.

12.5.3 SYSTEM IMPLEMENTATION

The series of events in the wireless home automation system is shown in Figure 12.2. Once the connection has been established, the parameter reading from various sensors like pa1, pa2, and pa3 has been noted. The threshold limits for the acceptable

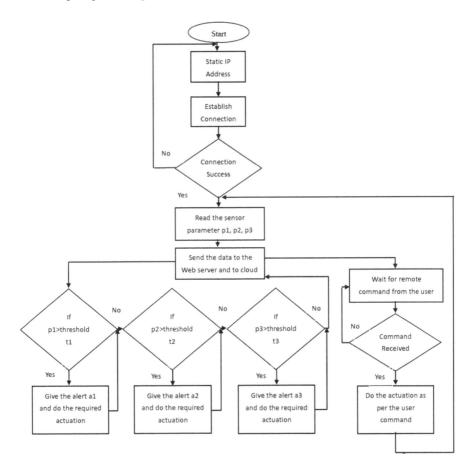

FIGURE 12.2 Flow of wireless home automation system.

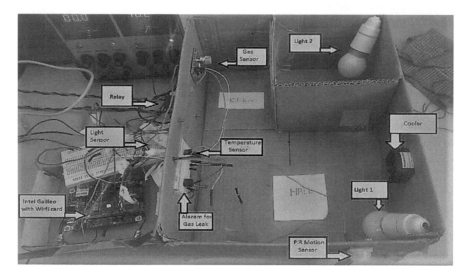

FIGURE 12.3   Prototype of home automation system.

sensors are set as th1, th2, th3, then on. Data collected from the sensors are shipped to the online server and stored within the cloud, and they are often accessed and analyzed from anywhere all 24×7. If the data obtained from the sensor are found beyond the threshold limit, an alert will be generated by the respective alarm a1, a2, a3, etc., and on the basis of the alarm, the parameters will be controlled.

Figure 12.3 shows the prototype that is used to detect if there is any activity happening near the door; a motion sensor is mounted at the entrance of the door. The light sensor is used to sense darkness, and when it is detected, it automatically switches on Light 1. A temperature sensor is fixed to detect room temperature, and when it exceeds the set threshold limit, it will automatically turn on the cooler/fan. An alarm will be raised if any gas leakage is detected by the MQ-6 gas sensor mounted in the kitchen. The electrical devices such as light and fan are operated with the aid of relay. The server located in the storeroom is connected via Wi-fi.

12.6   RESULTS AND DISCUSSION

Once the successful connection is established with the server, the information caught from different sensors is distributed to the server for further observing the system. Figure 12.4 depicts the online server page, which shows how the system is monitored and controlled. The online server page is often obtained by giving the assigned IP address within the browser. The online server provides the knowledge about the temperature inside the places of the house and motion detected near the door. It will also give the status of varied appliances, which we control remotely.

Figure 12.5 shows the temperature data collected at various time spans in degrees Celsius and furthermore shows the data detected by the motion sensor alongside the detection state and time and therefore the number of detected times. All of that information is stored within the cloud and may be accessed anywhere 24×7 by the user.

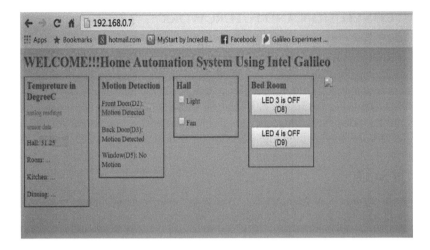

FIGURE 12.4   UI for server page.

	A	B	C	D
1		Temperature in DegreeC	Threshold Temperature	Motion Status
2	5/30/2015 16:54:42	30.23542	35	SAFE
3	5/30/2015 17:06:36	30.23542	35	SAFE
4	5/30/2015 17:06:43	30.23542	35	SAFE
5	5/30/2015 17:06:54	30.23542	35	SAFE
6	5/30/2015 17:07:05	31.56982	35	SAFE
7	5/30/2015 17:07:20	30.23542	35	SAFE
8	5/30/2015 17:07:28	30.23542	35	Motion Detected
9	5/30/2015 17:07:39	30.23542	35	Motion Detected
10	5/30/2015 17:07:39	32.22657	35	SAFE
11	5/30/2015 17:07:50	30.23542	35	SAFE
12	5/30/2015 17:07:50	30.23542	35	SAFE
13	5/30/2015 17:08:06	30.23542	35	SAFE
14	5/30/2015 17:08:06	32.22657	35	SAFE
15	5/30/2015 17:08:13	30.23542	35	SAFE
16	5/30/2015 17:08:13	30.23542	35	SAFE
17	5/30/2015 17:08:24	30.23542	35	SAFE
18	5/30/2015 17:08:24	30.23542	35	SAFE
19	5/30/2015 17:08:35	31.56982	35	SAFE
20	5/30/2015 17:08:36	32.22657	35	SAFE
21	5/30/2015 17:08:51	32.22657	35	SAFE
22	5/30/2015 17:08:51	32.22657	35	SAFE
23	5/30/2015 17:08:58	32.22657	35	SAFE
24	5/30/2015 17:08:58	32.22657	35	SAFE
25	5/30/2015 17:09:09	32.22657	35	SAFE

FIGURE 12.5   Temperature data stored in the cloud.

12.7 CONCLUSION AND FUTURE WORK

The proposed system has been experimentally proven to function effectively by controlling different appliances using the Web of Things, and it is designed to trace sensor data and operate the various appliances as required, like switching on light when darkness is detected. In the future, the system is often extended by adding security measures, capturing the image of a stranger moving around the house, and storing it within the cloud for further analysis if any unauthorized person is suspected. In terms of storage, this method might be simpler than employing a CCTV camera.

REFERENCES

[1] K. B. Sarmila, S. V. Manisekaran, "A study on security considerations in IoT environment and data protection methodologies for communication in cloud computing", in *Security Technology (ICCST) 2019 International Carnahan Conference*, pp. 1–6, 2019.

[2] J. Gubbi, R. Buyya, S. Marusic, M. Palaniswamia, "Internet of Things (IoT): A Vision, Architectural Elements, and Future Directions".

[3] S. P. Pande, P. Sen, "Review on: home automation system for disabled people using BCI," in *IOSR Journal of Computer Science (IOSR-JCE)* e- ISSN: 2278–0661, p-ISSN: 2278–8727, pp. 76–80.

[4] B. Hamed, "Design & implementation of smart house control using LabVIEW", *International Journal of Soft Computing and Engineering (IJSCE)* 1, 6, 2012.

[5] B. M. M. El-Basioni, S. M. A. Elkader, M. A. Fakhreldin, "Smart home design using wireless sensor network and biometric technologies", *Information Technology* 2, 3, 2013.

[6] I. Kaur, "Microcontroller based home automation system with security", *International Journal of Advanced Computer Science and Applications* 1, 6, 2010.

[7] R. J. Robles, T.-H. Kim, "Review: context aware tools for smart home development", *International Journal of Smart Home* 4, 1, 2010.

[8] K. Sujatha, S. Chidambaranathan, J. Janet, "Clever card novel authentication protocol (NAUP) in multi-computing internet of things environments", *International Journal of Pure and Applied Mathematics* 116, 12, 1311–8080 (printed version), 11–19, 2017.

[9] H. Rawat, A. Kushwah, K. Asthana, A. Shivhare, "LPG gas leakage detection & control system", in *National Conference on Synergetic Trends in engineering and Technology (STET-2014) International Journal of Engineering and Technical Research*, ISSN: 2121–0869, Special Issue.

[10] A. AlHammadi, A. AlZaabi, B. AlMarzooqi, S. AlNeyadi, Z. AlHashmi, M. Shatnawi, "Survey of IoT-based smart home approaches", in *Advances in Science and Engineering Technology International Conferences (ASET)*, pp. 1–6, 2019.

[11] D. Nicholas, B. Darrell, S. Somsak, "Home automation using cloud network and mobile devices", in *2012 Proceedings of IEEE Southeastcon*, IEEE, 2012.

[12] M. Chan, E. Campo, D. Esteve, J. Y. Fourniols, "Smart homes-current features and future perspectives", *Maturitas* 64, 2, 90–97, 2009.

[13] Z. Ahmad, M. H. Abbasi, A. Khan, I. S. Mall, M. F. Nadeem Khan, I. A. Sajjad, "Design of IoT embedded smart energy management system", in *Engineering and Emerging Technologies (ICEET) 2020 International Conference*, pp. 1–5, 2020.

[14] D. Javale, M. Mohsin, S. Nandanwar "Home automation and security system using android ADK", in *International Journal of Electronics Communication and Computer Technology (IJECCT)*, vol. 3, no. 2, 2013.

[15] A. R. Al-Ali, M. Al-Rousan, "Java-based home automation system", *IEEE Transactions on Consumer Electronics* 50, 2, 2004.

[16] U. Singh, M. A. Ansari, "Smart home automation system using internet of things", in *Power Energy Environment and Intelligent Control (PEEIC) 2019 2nd International Conference*, pp. 144–149, 2019.

[17] V. V. Kaveri, V. Meenakshi, B. Bharathi, J. Albert Mayan, "Smart garbage monitoring system using IOT", in *Advances in Computational and Bio-Engineering, Proceeding of the International Conference on Computational and Bio Engineering*, vol. 2, pp. 421–428, 2019.

[18] C. Perera, A. Zaslavsky, P. Christen, D. Georgakopoulos, "Context aware computing for the internet of things: a survey", *IEEE Communications Surveys & Tutorial* 16, 1, 414–454.

[19] J. Janet, S. Balakrishnan, E. Murali, "Improved data transfer scheduling and optimization as a service in cloud", in *2016 International Conference on Information Communication and Embedded Systems, ICICES 2016*, pp. 1–3, 2016.

[20] S. R. Das, S. Chita, N. Peterson, B. A. Shirazi, M. Bhadkamkar, "Home automation and security for mobile devices," in *IEEE PERCOM Workshops*, pp. 141–146, 2011.

[21] V. V. Kaveri, V. Maheswari. "A framework for recommending health-related topics based on topic modeling in conversational data (Twitter)", *Cluster Computing* 22, 5, 1–6, 2017.

[22] R. Piyare, "Internet of things: ubiquitous home control and monitoring system using android based smart phone", *International Journal of Internet of Things* 2, 1, 5–11, 2013. doi: 10.5923/j.ijit.20130201.02.

[23] G. Kortuem, F. Kawsar, D. Fitton, V. Sundramoorthy, "Smart objects as building blocks for the internet of things", *IEEE Internet Computing* 14, 44–51, 2010.

[24] V. R. Srividhya, S. A. Niketh, N. Kavarvizhv, "Information abstraction from IoT streaming greenhouse data", in *Recent Trends on Electronics Information Communication & Technology (RTEICT) 2019 4th International Conference*, pp. 235–241, 2019.

[25] C. Pereray, A. Zaslavskyy, P. Christen, D. Georgakopoulosy, Research School of Computer Science, The Australian National University, Canberra, ACT 0200, Australia CSIRO ICT Center, Canberra, ACT 2601, Australia CA4IOT: Context Awareness for Internet of Things.

[26] S. Hilton, Progression from M2M to the Internet of Things: An Introductory Blog. Available: http://blog.bosch-si.com/progression-from-m2m-tointernet-of-things-an-introductory- blog/, 2012.

[27] C.-H. Chen, C.-C. Gao, J.-J. Chen, "Intelligent home energy conservation system based on WSN," presented at the International Conference on Electrical, Electronics and Civil Engineering, Pattaya, 2011.

[28] "U.S. Patent 613809: Method of and apparatus for controlling mechanism of moving vessels and vehicles". United States Patent and Trademark Office. 1898-11-08. Retrieved 2010-06-16.

[29] W. C. Mann (ed.), *Smart Technology for Aging, Disability and Independence: The State of the Science*, John Wiley and Sons, 2005, pp. 34–66.

[30] K. S. Sankar Javvaji, U. R. Nelakuditi, B. P. Dadi, "IoT based cost effective home automation and security system", in *Computing Communication and Networking Technologies (ICCCNT) 2020 11th International Conference*, pp. 1–5, 2020.

13 IoT Platform for Monitoring and Optimization of the Public Parking System in Firebase

T. Kalavathi Devi
Kongu Engineering College

S. Umadevi
Vellore Institute of Technology

P. Sakthivel
Velalar College of Engineering and Technology

CONTENTS

13.1 INTRODUCTION AND LITERATURE SURVEY

The key objective of the proposed work is to increase the operational efficiency of current parking systems and also to naturally make the parking process more coordinated. The user is provided with a hassle-free Web application that gives information on the parking status earlier and acts as a user interface to identify and track locations. The uniqueness of this research work is that the parking process does not rely solely on one method of finding an appropriate slot. One possible technique is to use IR sensor itself to find the location, and based on mobile application, the data transfer is achievable.

DOI: 10.1201/9781003240853-13

Urbanization growth in international locations, which includes India and China, has precipitated huge bursts of the population of their metro cities like New Delhi and Beijing. The transportation network inside and across the cities plays an incredible role [1] in the socioeconomic development on state- and country-wise. In India, the urban population is on the rise at a typical annual pace of about 3%. This rise will impact the figure of 500 million out of the current 377 million by 2021 in urban areas [2], according to a survey conducted. The rapid rise in the city growth creates a demand for mobility and usage of vehicles individually on the roadways. Increase in vehicular movements increases the traffic particularly in city locations.

In step with the ministry of city development, Government of India, cities like Hyderabad, Chennai, and Kolkata metropolis record a higher percentage of parking in the street (30% roughly) when compared to Bengaluru, Delhi, and Mumbai. People in the region of Kolkata mostly favor individual vehicles that they utilize most of the parking slots with extensive time consumption, whereas in the other cities, common transportation is typically preferred. Thus, the usage of public vehicles involves the diminution of sharing ratio of parking vehicles in the parking slots. But on considering the street parking index reference, it is in increasing order for the cities like Delhi, Kolkata, and Mumbai when compared to metropolitans like Hyderabad, Chennai, and Pune. Due to the high parking charge collection at the common parking centers, people prefer to park on the streets without considering the warnings issued by the governments. Figure 13.1 indicates the parking status of the vehicle on the roadside. This is one of the common reasons for the necessity to enhance the public parking slots.

An intelligent car parking [4] system is based on wireless sensor networks with the arrangement of low-cost sensors operating in a wireless mode that is placed in the car parking area. The exploitation facility available in the parking location is tracked by the sensor node available in the slot. The active photoelectric sensors were

FIGURE 13.1   Parking status on the roadside (courtesy Google image).

kept at the parking entry/exit along with the centralized server for proper sensing and propagation of necessary information. This pact provides information on the list of parking slots, which is free without giving them information on the exact location. This increase in parking slots creates an exponential issue in getting the exact free parking slot. Ultrasonic sensor is to locate parking garage [6] habitat or unsafe parking. Irrespective of the minimal effort and the simple installation of ultrasonic sensors, the parking system may receive incorrect data due to the sensor's sensitivity to temperature changes and extreme air turbulence. Multi-level car parking system with image processing technique [7] has obtained information on the existing parking spaces, process them, and then place the vehicle at that position. In this approach, the cameras placed at the entrance of the parking slots help to gather information about the vehicle and the user, rather than incorporating additional components or algorithms. Thereby the image processing algorithm presented in the above research work permits several vehicles for processing with a single camera. This method further also contributes to the configuration of another version of the structure that focuses on the reception and processing of the captured images. A common technique for this situation is using wide-angle cameras collectively with a category set of rules to stumble on parking lines and to determine the frame area for the vehicle as in [8]. Besides, a method that involves facing camera on the rear side is addressed in [9]. It assists the drivers to park in an allotted spot with the help of a touch screen. This method consists of three steps: the creation of a bird-eye view, recognition of a guideline, and identification of parking separating line. The accuracy of the captured images is enhanced by increasing the training phase of the algorithm with various images obtained at different climatic circumstances [10]. The process is that the database of the area occupied by the parking slot along with the free space is stored in the memory and the user information is compared with the stored database; the decision is furnished to the user. Another approach within the camera-based detection is utilizing the camera facility inbuilt within the car drivers center [11–13] mirror itself for determining the free parking slot. One of the major troubles our country ahead's with an increase in urbanization and development is the usage of private vehicles. People have begun using their means [14–17] of transport to move from one place to another, causing traffic congestion and pollution. The use of private vehicles has raised the issue of public parking. This has become one of the most inevitable challenges people face in their daily lives. Public parking is a time-consuming operation. Searching for an acceptable parking spot may be challenging, and they may have to travel around a lot that causes traffic congestion, air pollution, and time loss. Having experienced the annoyance of needlessly circling a big one in search of a parking space, it is worth having a parking slot in advance. This will not only alleviate the issue of parking but will also allow the owner of the vehicle to drive easily, thereby reducing his/her difficulty in finding a parking space. Internet of Things (IoT)-based traffic violation and smart home automation designs have been conversed in [16–20].

The proposed automated parking system helps the vehicle owner to find the empty slots even before parking. With the help of mobile application, the user has to book the required details for parking the car in the slot. The database of the user is maintained in the open-source software Firebase. Based on the slot availability, the owner gives the decision for booking the slot. The IoT-based data maintenance permits the automobiles datum to be properly continue or remotely access the prevailing network infrastructure, by devising chances for direct integration of the real-world data into computer-based systems. This results in an increased efficiency, accuracy, and economic merits with the ancillary benefits of the minor reduction in searching for a parking place for the concerned owner/driver of the vehicle. In the parking slot if no slots are available, the servo motor will not open the gate, and the message space "not available" will be displayed on the monitor in front of the car parking gate. This is possible with the use of different hardware and software. On the hardware side, the base of the project is Arduino Uno Board and Wi-fi module which is a user-friendly device. They can be easily interfaced virtually with any sensors or modules on the market today. While Arduino core is used on the software side to run the Arduino board, MySQL is used to maintain and track all records in its database, and PHP is used to retrieve all such data. At last, vehicle's smart parking system operates by tracking the availability of car parking spaces and making the information available to drivers and facility administrators. In real time, the information is sent to intelligent cellular devices with the use of wireless networks. Ultimately, the automobile's smart parking device operates by way of monitoring the availability of vehicle parking areas and making the statistics available to drivers and facility directors.

The chapter is organized as follows: Section 13.1 discusses the introduction and survey of the existing works. Section 13.2 describes the methodology and software used. Hardware implementation is given in Section 13.3. Section 13.4 addresses the results and comparison of the proposed work.

13.2 METHODOLOGY

The architecture of parking management systems with IoT is shown in Figure 13.2. The sensing block consists of infrared sensors, ultrasonic sensors with the communication to the network layer by gateways, and IoT data server ThingSpeak. Data processing layer has portable Arduino controller and standard Wi-fi module. Power delivery to the independent circuit is promptly done with a 12 V transformer. The input voltage required by the Arduino microcontroller is driven by the regulated voltage. The voltage regulator (5 V) is placed to regulate the 5 V input required for the Arduino microcontroller. The infrared sensors sense the presence or absence of a car in the parking slot.

The flow diagram and circuit diagram of the parking management system are given in Figure 13.3a and b. The parking system is incorporated with IR sensors that sense the presence of the vehicle. If the vehicle is present in the slot, it is updated in the application through the database. Here, the IR sensors are connected to the power supply of 5 V through Arduino. In a parking slot, the availability of car is detected by the IR sensor, which works on the following criteria: When there is a car ahead of IR sensing element, the IR transmitter waves from the car are received by the IR receiver. At this state, the condition of the IR is zero. Similarly, if there is no vehicle in the slot, then there will not be any transmitted wave and the output is 1. The Arduino receives the sensed inputs from the IR sensors and gives them to the Wi-fi module to update the respective state of the parking slots to the database. ATmega328P efficiently performs the key role of portable controller in this specific project. ATmega328P is a

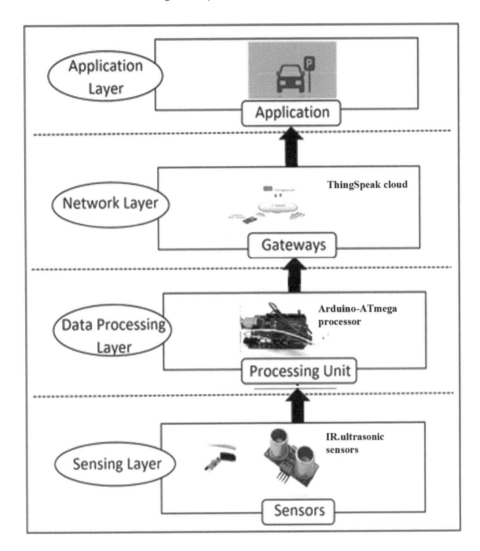

FIGURE 13.2 Architecture of the parking management system.

tiny microchip coded using embedded C language. It typically has both analog and digital input and output. It is a low-power complementary metal oxide semiconductor (CMOS) eight-bit microcontroller.

In a transformer, the direct conversion of 230 V AC voltage is converted into 12 V DC voltage which can be used by Arduino board to power up the board. The voltage regulator is used to maintain the constant voltage for the IR sensors. The Wi-fi module used in the Arduino board is ESP8266 NodeMCU, which uses a 3.3 V supply, and this helps to perform the IoT interface with the server. This Wi-fi module is an independent system on chip with TCP/IP protocol to provide wireless access for the Arduino controller. The interfacing cloud used is the ThingSpeak by creating a channel, and the information is transmitted from the sensor to the storage server. Firebase

(a)

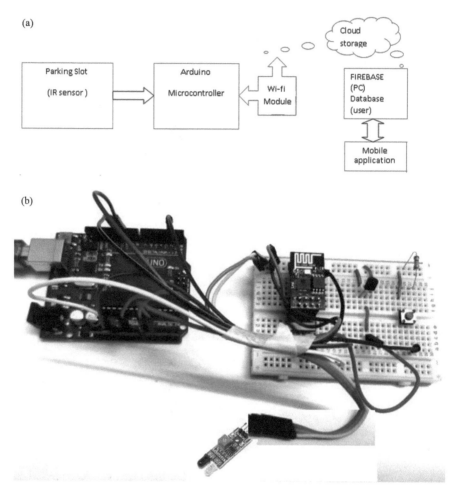

(b)

FIGURE 13.3   (a) Flow diagram of the parking management system. (b) Circuit diagram of the parking management system.

is the open-source software used to maintain the database of the users. Based on the information from the sensor, decision will be taken by the owner.

13.3 IMPLEMENTATION

Firebase is a cellular podium that smoothes the progress of quickly building up high-quality applications, generates a user base, and gets monetary benefits. Firebase is made from complementary capabilities that can blend and match to fit the wishes, with Google Analytics for Firebase on the center. Firebase offers a real-time database and back end as a service provider. This service presents the application developers with an API that permits utility statistics to be synchronized across clients and saved on Firebase's cloud. The mobile application interacts with the end user for verifying the availability of the channels, for booking the slots, and the user can prefer the most neighboring parking area. The specific output from the IR sensors is instantly updated in the centralized database whether the slot is vacant or not. The user details of booking the slots in the mobile application are filled in the created database. The home page of the application has options for the public and the parking owner. The community who wants to book a parking slot in the desired area has to log in as the public. The owner also can use this mobile application for accepting the slots booked by the user; accept the request by selecting the specified parking slot and typing the password known only to the owner. The users can login by entering their registered mobile number. The new users who have not registered their mobile number can register on this page by giving their name, email id, and phone number. On clicking the parking area, it shows the number of parking slots and the status of the parking slots. The parking slots available to the nearby location are displayed on the screen. Based on the available parking slots, for example, if slot 1 is booked and the vehicle is parked in slot 3, then they cannot be booked. Slot 2 can be booked. After selecting the desired slot, a timer starts for 3 minutes and a request is sent to the authority. If no response is received within 3 minutes, then the slot will be un-booked automatically. On the other side, if the user has booked the slot prior but he could not be able to utilize the slot, then a reminder message is sent to their registered mobile number. It waits for 3 minutes, and automatically, the booking gets cancelled.

The next page of the parking owner option is the selection of the parking area by the user. The authorities at the respective locations can access their parking area by choosing the parking area and entering the passcode. If the authority allows, the user will be provided with the parking information and he/she may park the vehicle. If the authority rejects the request, then it will go to the home page. This page also shows the logistics of the users who previously booked the slots. The entire screen of the parking details is given in Figure 13.4.

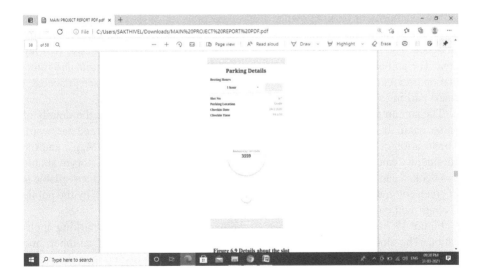

FIGURE 13.4   Screen of the parking details of the mobile application.

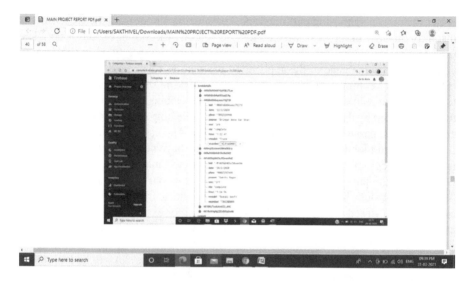

FIGURE 13.5   Developed database using open-source software Firebase.

The details of the users booking the parking slot through the mobile application are stored in the database, which is developed using open-source software Firebase [21] and given in Figure 13.5.

Figure 13.6 gives the entire owner login credentials. The user details like phone number, vehicle number, and the car model type are fetched in the application, while booking the parking slot is stored in the database so that it could be analyzed by the parking authority or owner. The sensor updating in the Firebase shows the car park status of the IR sensor installed in the parking area. The sensor status is also monitored in the database for further reference. The park login is for access of the parking authority or parking owner. It has passcodes for different parking locations using which the authority can access the parking location by entering the respective passcode. The users' requests for parking will be displayed in the respective parking area. The authority can allow or deny the users' request as per the details and information provided. The mobile application has to be installed by the user to access a smart parking system.

The signup page is opened, and the user can register by entering the details of name, email id, and phone number and then log in using the user name and password. The location of the customer is made visible by linking with Google Maps, and the community can choose the desired parking area which is nearest. The slots available and booked can be viewed by clicking the desired parking area. If the slot available is in green color, it is not booked; otherwise, it is booked. The slot chosen by the public is confirmed only when the parking owner accepts the request.

Figure 13.7 indicates the flowchart of the parking owner using the parking application. The slot is booked with the car details including car model, number, and phone number. If the slot request is accepted by the parking owner, then the public can view the slot for choosing the hours of the slot in the parking application. The public can use the checkout option when they completed the task of parking.

FIGURE 13.6   Owner login credentials in Firebase.

13.4   RESULTS AND DISCUSSION

The IoT implementation of the smart parking management system is shown in Figure 13.8. From Figure 13.8, it is found that the database of the user for parking is taken into the cloud during peak hours of initial value to 1 p.m., and from 1 p.m. to 3 p.m., the parking slot is found to be empty. Then, after 3 p.m., it goes to peak booking. In the next field chart, the server maintains the data point, which indicates that the booking of the slot is gradually increased, and beyond 3 p.m., there is traffic in slot booking. In the third field, it indicates that no parking slot is booked. Integrated data point in the centralized server is indicated in the image, and the database of the individual is given there. This information is relevant for the user to carry out decisions when they move out of the home. It consists precisely of the four blocks, which show the user information's feeding into the mobile application, data collected in the server, the database of the user's information with the help of Arduino Uno, NodeMCU, and logic level converter. Necessary IoT infrastructure is properly established using ESP8266 to enable and carefully collect and adequately develop the data with no difficulty.

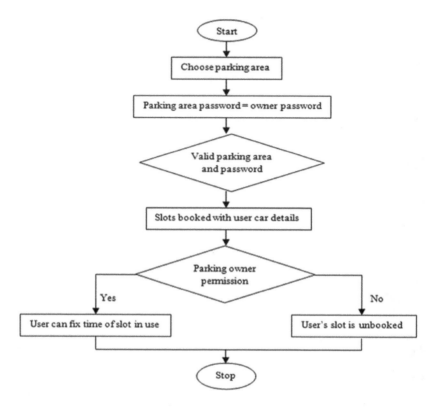

FIGURE 13.7 Flowchart of parking owner using the parking application.

FIGURE 13.8 IoT database of the parking management system.

This module can be integrated with the ThingSpeak site to monitor the data through the Internet. It collects the data from the sensors connected to the Arduino. IoT communications produce source data from the conditions of the parking slots (i.e., the availability of the parking slot) of the parking places within SC. The proposed smart parking system uses the collected data from parking lots parking slots along with the data collected at the particular location and instant of the vehicle. The setup of the parking slot has the IR sensors placed in the parking slots. The parking location is given by the mobile application, and if a vehicle is booked for parking, the sensors detect and information is stored and updated in the application through the database.

Firebase is reliably found to constitute an effective open source for implementing the official database when compared to OSS like Back4App, Parse, AWS Amplify, Kuzzle, and Hoodie. It is suitable for cloud-enabled real-time data storing purposes, which is used with Android and iOS, so that the clients can share the database instantly in bidirectional mode. Firebase is compared with PostgreSQL, MySQL, SQLite, and MariaDB [22]. Figure 13.9 shows the parameters that are taken for comparison. The execution time vs number of records using ten client details is compared. These databases make use of all the cores of the test system and describe how increasing the number of records affects the average response time of a database operation. In comparison, the user information is sent to the Firebase with all details like name, email id, phone number, and car details, and execution time required for the data is in the order of 0.5–25 ms when compared to the other databases. Also, the important advantage is that information of the users is automatically updated without any delay. This is powered by Google so that the location tracking to find the parking slot is quite comfortable. There is no need for foreign keys with good transaction concepts. The concurrency nature of the database and robustness in the coding claims are to be better among the other sources.

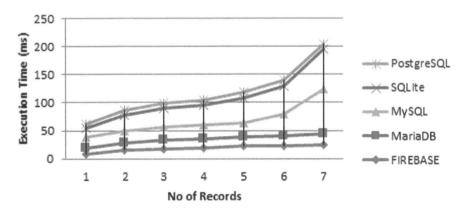

FIGURE 13.9   Performance comparison of Firebase.

13.5 CONCLUSION

It is hard to explain the difficulty of the driver to identify the parking slot in busy timing schedules. If no parking slot is found, then parking of the vehicles on the no parking spaces and the roadside is getting too busy. This method discusses the IoT-based parking system along with the database created using open-source software like Firebase. The controller involved in the decision-making is the Arduino with IR sensors. The cloud information is taken care of by the node ESP8266 that is inter-faced with Arduino. With the invaluable help of the ThingSpeak cloud, the personal data are stored in the server and updated automatically to retrieve the information of the user in the database or the owner of the parking space. The results indicate that almost 50 users have utilized the application-based registration and the graphs obtained from the cloud indicate the information and time retrieval about the users. The execution time required for the data is in the order of 0.5–25 ms when compared to the other databases, which is very minimum. The work has the advantage of using the mobile application with Firebase and IoT-based data maintenance.

REFERENCES

[1] S.S. Thorat, M. Ashwini, A. Kelshikar, S. Londhe, M. Choudhary, (2017). IOT Based Smart Parking System Using RFID, *International Journal of Computer Engineering In Research Trends*, 4(1), 9–12.

[2] S.K. Singh, (2012). Urban Transport in India: Issues Challenges and the way Forward, *European Transport*, 52, 27–52. https://www.researchgate.net/publication/286966258_Urban_transport_in_India_Issues_challenges_and_the_way_forward/citation/download.

[3] G. Kodransky Hermann, (2017). Europe's Parking U-Turn: From Accommodation to Regulation, Institute for Transportation and Development Policy, September 15, 2017. https://itdpdotorg.wpengine.com/wpcontent/uploads/2014/07/Europes_Parking_U-Turn_ITDP.pdf

[4] M. Madhan Mohan, A. Akshaya, E. Anjali, A. Dharisini, V. Harish, (2020). Development of Spinning Monitor to Facilitate the Sider in Textile Mills, *International Journal of Recent Technology and Engineering (IJRTE)*, 9(1), 981–985. doi:10.35940/ijrte.A2167.059120.

[5] V.W.S. Tang, Y. Zheng, J. Cao, (2006). An Intelligent Car Park Management System Based On Wireless Sensor Networks, *IEEE SPCA06: 1st International Symposium on Pervasive Computing and Applications*, August 3–5, Urumchi, Xinjiang, P.R. China, 65–70. doi: 10.1109/SPCA.2006.297498.

[6] J. Vera-Gómez, A. Quesada-Arencibia, C. García, R. Suárez Moreno and F. Guerra Hernández, (2016). An Intelligent Parking Management System for Urban Areas, *Sensors*, 16(6), 931–1–16. doi: 10.3390/s16060931.

[7] A. Kianpisheh, N. Mustaffa, P. Limtraiut, P. Keikhosrokian, (2014). smart Parking System (SPS) architecture using ultrasonic detector, *International Journal of Software Engineering and Research (IJSER)*, 2(6), 21–26. https://www.researchgate.net/publication/230701092_Smart_Parking_System_ SPS_Architecture_ Using_ Ultrasonic_Detector/citation/download.

[8] M.I. Reza, A. Rokoni, M. Sarkar, Smart parking system with image processing. *Intelligent Systems and Applications*, 4(3), 41–47. https://www.irjet.net/archives/V5/i4/IRJET-V5I4201.pdf.

[9] S.E. Shih, W.H. Tsai, (2014). A convenient vision-based system for automatic detection of parking Spaces in indoor parking using wide-angle cameras, *IEEE Transactions on Vehicular Technology*, 1–12. doi: 10.1109/TVT.2013.2297331.

[10] R. Grodi, D.B. Rawat, F. Rios-Gutierrez, (2016). Smart parking: Parking occupancy monitoring and visualization system for smart cities, In *Proceeding of the southeast conferences*, Norfolk, VA, 1–5. doi: 10.1109/SECON.2016.7506721.

[11] L. Baroffio, L. Bondi, M. Cesana, E. Redondi, M. Tagliasacchi, (2016). A visual sensor network for parking lot occupancy detection in Smart Cities, In *Proceedings of the IEEE World Forum on Internet of Things*, December 2016, 745–750. doi: 10.1109/WF-IoT.2015.7389147.

[12] C.F. Yang Ju, Y.H. Hsieh, C.Y. Lin, M.H. Tsai, H.L. Chang, (2017). parking—A real-time parking space monitoring and guiding system, *Vehicular Communications*, 9, 301–305. doi:10.1016/j.vehcom.2017.04.001.

[13] T. Kalavathi Devi, P. Sakthivel, (2020). Low power sleepy keeper technique based VLSI architecture of Viterbi decoder in WLANs, *Australian Journal of Electrical and Electronics Engineering*, 7(4), 263–268, doi:10.1080/1448837X.2020.1844366.

[14] B. Karunamoorthy, R. SureshKumar, N. JayaSudha, (2015). Design and implementation of an intelligent parking management system using image processing, *International Journal of Advanced Research in Computer Engineering Technology*, 4, 85–90. http://ijarcet.org/wp-content/uploads/IJARCET-VOL-4-ISSUE-1-85-90.pdf.

[15] S. Lin, Y. Chen, S. Liu, (2006). A vision-based parking lot management system, In *Proceedings of the IEEE International Conference on Systems, Man, and Cybernetics*, Taipei, 8–11. doi: 10.1109/ICSMC.2006.385314.

[16] K.D. Thangavel, S. Palaniappan et al., Analysis of overloading in trucks using embedded controller, In *Proceedings of the International Conference on Electronics and Sustainable Communication Systems (ICESC 2020)* IEEE Xplore Part Number: CFP20V66-ART, 944–949.

[17] A. Mulyun, D. Parikseit, (2010). Analysis of loss cost of road pavement distress due to overloading freight transportation, *Journal of the Eastern Asia Society for Transportation Studies*, 8 90–98. doi:10.11175/easts.8.706.

[18] C. Aravind, S.J. Suji Prasad, M. Ponnibala, (2020). Remote Monitoring and control of automation system with the Internet of Things, *International Journal of Scientific and Technology Research*, 1(1), 945–948. http://www.ijstr.org/final-print/jan2020/Remote-Monitoring-And-Control-Of-Automation-System-With-Internet-OfThings.pdf.

[19] R.S. Guru Prasad, K.N. Baluprithviraj, S. Idhikash, S.R. Kirubaharan, V. Ashwin, (2020). Automatic penalty of vehicles for violation of traffic rules using IoT, *International Journal of Recent Technology and Engineering (IJRTE)*, 8(6), 415–420. doi:10.35940/ijrte.E5621.038620.

[20] V. Goyal, K.S. Umadevi, (2017). Simple architecture for smart home using Internet of Things, In *International Conference on Innovations in Power and Advanced Computing Technologies*. doi: 10.1109/IPACT.2017.8245145.

[21] Am-suk oh, (2021). Design and Implementation of Platform for Monitoring of Notification System in FIREBASE Message, *Journal of Information and Communication Convergence Engineering*, 19(1), 16–21, 2021, doi: 10.6109/jicce.2021.19.1.16.

[22] S. Adeyi, Y. Abubakar, A.A. Oriyomi, (2014). Benchmarking popular open source RDBMS: A performance evaluation for IT professionals, *International Journal of Advanced Computer Technology (IJACT)*, 3(5), 39–45.

14 IoT-Based Smart Health Monitoring System

*Priyanka Lal, Ananthakrishnan V,
and O.V. Gnana Swathika*
Vellore Institute of Technology

Naveen Kumar Gutha
Andhra Loyola Institute of Engineering and Technology

V Berlin Hency
Vellore Institute of Technology

CONTENTS

14.1 INTRODUCTION

14.1.1 OBJECTIVE OF THE PROJECT

Health monitoring is one of the dominating problems all over the world in spite of the fact that health plays a unique role in a person's overall development and growth.

Importance of health cannot be neglected at any cost as it is a substantial part of a person's mental or physical productivity. But, due to lack of proper health monitoring or health awareness, patients had to suffer from serious health issues that could have been eradicated by providing proper health service/care at the right time. Elderly people are the most affected section of the society because old age invites many health problems owing to the reduced immunity and weakened body. To overcome this problem, health experts are making use of technological advancements done in the field of medical sector. Today, technology has completely revolutionized the healthcare industry. Our objective is to design a health monitoring gadget using Internet of Things (IoT) that can be used for recording major factors affecting a person's health like heartbeat rate (HR), body temperature (BT), and detect free fall, which can be further processed to observe a person's health. Gathered information can be sent to that person's relatives or doctor.

14.1.2 SCOPE OF THE PROJECT

The IoT has empowered medicinal services observing to turn out to be progressively boundless and successful. Previously, patients must be observed in a clinical office or under the consideration of family or home medical caretakers. In the event that a patient chose to mend in a medical clinic, their imperative signs – pulse, glucose levels, and heart levels – could be observed by medicinal service experts. Be that as it may, if a patient chose to recuperate at home under the watchful eye of family, they gambled not having the option to promptly distinguish inconveniences from ailment and illness.

With IoT advancements facilitating the ease of remote patient observation, patients no longer need to pick between living autonomously and having a sense of security should. With the predictable checking given by IoT advances and continuous alarms, patients and their family have a suspicion that all is well and good regardless of whether the patient chooses to be at home. As indicated by Grand View Research, the worldwide IoT remote well-being observing business sector is required to develop from \$58.4 billion on 2014 to more than \$300 billion by 2022. Later, IoT well-being checking will give expanded autonomy and portability to old, wiped out, and truly or intellectually handicapped patients and lessen worry for family by alarming the specialists who can respond quickly when issues emerge.

14.1.3 BACKGROUND

The leading edge of baby boomer (those people born worldwide between 1946 and 1964) age began turning 65 out of 2011. In India, more than 10,000 people born a day after World War II (WW II) are turning 65 and their number is expected to be doubled by the year 2050, to 80 million owing to the constant growth of the senior – and the vast majority of that growth began in 2010 and will proceed until 2030, when that generation of baby boomers will enter their senior years. The US Census Bureau's prediction says that an average growth 2.8% per year will be seen in the number of seniors during that time. In Canada, a fourth of the populace by 2041 will be comprised of seniors. So, this kind of market, which deals with age and growth, is only going to get bigger. Thus, for people who are willing to meet their needs, it can provide even more business opportunities. People aged 65 or above usually live at home, with their respective spouse or alone. A research conducted by the American Association of Retired Persons (AARP) found that about 90% of the senior citizens want to stay in their own homes or communities and thus wish to live independently, similar to young age people. Unfortunately, old age is an invitation to many diseases, aches, and disabilities. This necessitates the compulsory requirement of seeking assistance and medical help from health experts as mostly elderly people having an activity restriction or a disability are residing alone. Many chronic health problems including various serious disabilities are found on a wide scale among elderly people, every four out of five old persons staying at home have a chronic health problem of some kind. The most common of these conditions are (nonarthritic) back pain, arthritis or rheumatism, cataracts, hypertension, and heart diseases. This instills a sense of fear in the minds of seniors that if they fall due to cardiac arrest or other medical problem, then they may need to face some serious consequences like permanent physical damage of an organ or delay in recovery as they lack assistance. This can result in depression, lack of mobility, or some of the other serious conditions that can put the patient in a downward spiral. So here, the devices which are based on IoT-based smart healthcare system provide a solution for that by developing a smart health monitoring watch to help people carry on their day-to-day activities more independently and confidently, thus gifting them a happier and better life.

14.1.4 MOTIVATION

The elderly population of India is one of the fastest growing populaces in this whole world. At about 110 million, India has the second biggest worldwide populaces of aging citizens. By 2050, this number will most likely increment to 240 million. The growth of population aged above 65 years with respect to the total population can be analyzed in Figure 14.1. It can be clearly seen that the populace of elderly people has been increasing steadily with each decade from 1900 to 2010, thus increasing their percentage with respect to the total population by 2010 to an approximate percentage of 12.5%.

In spite of this huge population of senior citizens, India comes up short on the essential framework and medical experts to ensure the well-being of senior citizens. India does not have the basic infrastructure and medical support to take care of their health. As per numerous studies and surveys across the country, the greatest concerns for most of the Indian senior residents are cost of medicinal services, financial strain due to isolation, and lack of support. Furthermore, the greater part of the elderly residents is not agreed upon the poise of care they deserve. Absence of visible infrastructure is a significant obstacle in comfort to the elderly. There are not many consideration homes (old-age care homes) for them or even public ramps for the less versatile residents, similar to the individuals who need wheelchair access. With expanding life span and crippling ceaseless health conditions, many senior residents will require better access to physical framework in the upcoming years. This will be in both their own homes and broad daylight spaces, similar to streets and shopping centers. Very little data are available about the explicit geriatric illnesses. Emotional

Population 65 Years and Older by Size and Percent of Total Population: 1900 to 2010
(For more information on confidentiality protection, nonsampling error, and definitions, see *www.census.gov /prod/cen2010/doc/sf1.pdf*)

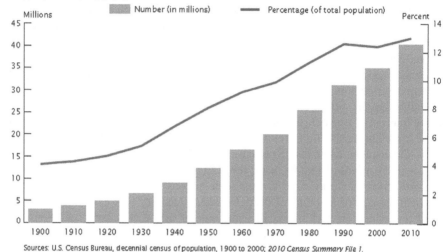

Sources: U.S. Census Bureau, decennial census of population, 1900 to 2000; *2010 Census Summary File 1.*

FIGURE 14.1 Increase in the population of people aged 65 years and above from 1900 to 2010.

wellness issues are infrequently discussed or examined, and thus, the nation is not well arranged to manage the expanding rate of dementia, Alzheimer's disease, and depression among the old. There are barely any offices and specialists who can adequately manage geriatric well-being, even in significant metros.

In contrast to many developed nations, emergency response framework for senior residents is not well designed in India, including the accessibility of ambulances for public at healthcare centers and hospitals. Perhaps, the greatest dread for elderly people living alone is how to approach and access an emergency facility whenever required, particularly at nighttime. Rapid socioeconomic changes, including progressive increase in nuclear families, is likewise making the care and management of seniors more troublesome, particularly for the grown-up youngsters, busy with their personal day-to-day life who are liable for their guardians' welfare and prosperity. Overseeing home care for the seniors is a major challenge as numerous service providers, who rarely converse with each other, are engaged in providing that care. These incorporate nursing organizations, physiotherapists, and clinical suppliers. Most of these providers are small, disordered, and low-maintenance players. Most senior residents who live alone endure from the absence of companionship.

As a significant part of our lives, the Internet has empowered enormous number of devices, machines, and gadgets we use in our daily lives to be continuously monitored and remotely controlled by making use of the technology called as IoT. Rapid advancements in this technology have transformed smart health applications into a quickly developing part of the health sector. For people suffering from cardiovascular diseases, the values of heart rate variability (HRV), HR, and BT are viewed as important determinants which should be measured on a regular basis. In a study, an application-based software is created that can record and analyze HRV, HR, and BT values for cardiovascular patients who ought to be under continuous monitoring. The wearable measuring device, which comprises various sensors and transducers, can record and analyze patients' symptoms constantly. After measuring, it sends the recorded data signals to all the registered devices by a wireless medium. If the measured values exceed the foreordained limit set for the patient, then the estimated values of HRV, HR, and BT are sent to the family members of the patient as well as to the specialist/doctor via e-mail and cell phone notification. Further, this device is also incorporated with fall detection technique and a panic button that immediately alerts the keens by sending a message via mail. This smart wearable measurement system permits a patient to be able to move freely in his/her own social surroundings, thus gifting them a confidence to carry on with their carefree lives.

14.2 LITERATURE REVIEW

All the references cited in this section are listed in Table 14.1. IoT-based smart healthcare system is proposed by Islam et al. [1], which is capable of monitoring a patient's fundamental well-being signs just as the room condition where the patients are present throughout the day. In this framework, five sensors are utilized to collect the information from patient's surroundings named heartbeat sensor, BT and room temperature sensor, CO_2 sensor, and CO sensor. For all cases, the probability of producing inaccurate results of the proposed framework is inside the limit of 0.05, means less than 5%. The state of the patients is communicated to hospital staff by means of an online gateway, where they can review and monitor the current condition of the patients. The created model is appropriate for easy observation and analysis of patient's health by medical institutions that is demonstrated by the viability of the system.

The technology focused in the paper by Al-khafajiy et al. [2] centers around the capacity to follow an individual's physiological information to identify explicit disorders that may help in early intervention practices. This is accomplished by carefully filtering and properly investigating the procured data of sensors while analyzing the identification of a disease/disorder to a relevant career. The finding uncovers that the proposed framework can help in improving clinical choice backings while encouraging early intervention practices. Results of our broad simulations indicate a high-level exhibition of the proposed framework. In this way, the framework runs proficiently and is financially savvy as far as collection and controlling of data are concerned. The ambient assisted living (AAL) programs/systems, generally, target at providing a carefree and secure living to old, disabled, and individuals with chronic problems in their homes. Moreover, AAL frameworks provide the ability to collaboratively and cooperatively assist a facility/condition managed by healthcare providers (e.g., relatives, companions, and clinical staff).

The paper by Tsukiyama [3] proposed a health monitoring system designed by using the sensors that assess the well-being status of single–independent old people by evaluating their day-to-day living habits and gives details of emergency forecasts to a nearby health facility/center without involving unnecessary client interaction.

TABLE 14.1

List of All the References Cited in the Literature Survey (Section 14.2)

Serial. No.	Citations
[1]	Islam et al. (2020)
[2]	Al-khafajiy et al. (2019)
[3]	Tsukiyama (2015)
[4]	Bourouis et al. (2011)
[5]	Bhadoria and Gupta (2013)
[6]	Park et al. (2017)
[7]	Shahriyar et al. (2009)
[8]	Sannino and Pietro (2011)
[9]	Huang et al. (2009)
[10]	Lv et al. (2010)
[11]	Gahlot et al. (2019)
[12]	Wan et al. (2018)
[13]	Dhakal et al. (2020)
[14]	Durán-Vega et al. (2019)
[15]	Wu et al. (2013)
[16]	Alnosayan et al. (2014)
[17]	Zhang (2019)
[18]	Appelboom et al. (2014)
[19]	Binti Abu Bakar et al. (2016)
[20]	Kumar et al. (2017)
[21]	Mahadewaswamy and Durgad (2018)
[22]	Rahman et al. (2020)
[23]	Mulla and Mote (2016)
[24]	Mahajan and Birajdar (2019)
[25]	Sharma et al. (2018)
[26]	Rahman et al. (2019)
[27]	Bangera and Dhiman (2017)
[28]	Siam et al. (2019)
[29]	Ali et al. (2018)
[30]	Samik et al. (2020)

Three everyday activities are primarily analyzed: kitchen activities, urination, and habits identified with keeping up physical hygiene in light of the fact that these exercises are firmly associated with keeping up a good health and are normally associated with the utilization of faucet water. There is a sure normality in the use of faucet water for these quotidian activities, and well-being status being associated with the regular habits as people pee a normal of six to eight or ten times a day. In this manner, if an individual makes significantly less or considerably large utilization of the washroom, at that point some disease or issue can be doubted. In the event that one wakes at least multiple times before daybreak to pee, it can likewise be an indication of a bad health condition. Consequently, the frequency of times of day that an individual pee is significant the indication of well-being status. It is conceivable to gather the well-being status by examining the progression of water at the latrine's flush tank in the washroom. Thus, this framework screens well-being status by utilizing water flow sensors, connected to fixtures in the kitchen and toilet/washroom. A bit of leeway of this strategy is that the framework can be promptly introduced in a lodging requiring little to no effort. Moreover, this system does not need individual's information to be spared or transferred. Underlying outcomes of the trial are also mentioned in the paper by Tsukiyama [3].

Ongoing researches and explorations in ubiquitous computation provides easy analysis of an individual's kinematics and physiological boundaries by utilizing advancements of body area networks (BANs). An ongoing portable well-being framework was proposed in the paper by Bourouis et al. [4], for checking old people from indoor to open-air conditions. The framework involves a smart phone as a focal hub with a biosignal sensor worn by the patient. The information is gathered from sensor and sent to a smart-integrated server via GPRS/UMTS for analysis. The model ubiquitous mobile health monitoring system (UMHMSE) screens the old person's versatility, environment, and fundamental signs, for example, HR and Sp02. Remote clients (also relatives or health experts) can get controls to view and monitor the gathered data through a Web application.

Portable registering [5] makes it conceivable to accomplish arrangement of individual social insurance and crisis caution and following which can screen individual well-being status in a continuous way and naturally issue the alarm for clinical guides if there should arise an occurrence of crisis by the following client's area accurately. With the advancement of the advances, for example, portable processing dispersed figuring and remote sensor organize, it is conceivable to furnish the older with medicinal service benefits that can screen the old whenever any place. This paper presents an audit on portable-based well-being checking frameworks. It likewise examines the alluring properties of a decent versatile social insurance framework just as hardly any current medicinal service applications are additionally quickly depicted.

The paper by Park et al. [6] explored the IoT human service innovation and the utilizations of enormous information. The upgrade of collaborations among advances and the remote social assurance has made quick improvements in IoT advancements, large information investigation, and wearable gadgets and in this manner produces big open doors for the mindful medicinal service applications. Wearable gadgets clear the way to remote medicinal service frameworks by clearing approach to information collection through cell phones. Constant handling of such crude information leads to the development of exploration zone, and information cleaning, information repreparing, information mining, and capacity of enormous informational collections are the results of the large information examination. This paper examines the need and the intensity of IoT-based remote medicinal service arrangements and their difficulties.

Depending upon the biomedical and surrounding information gathered using conveyed sensors, the intelligent mobile health monitoring system (IMHMS) demonstrated in the paper by Shahriyar et al. [7] is capable of giving appropriate clinical inputs or instructions to the patients via cell phones. As cell phones have become an indistinguishable piece of our life, it can coordinate social insurance all the more consistently to our regular daily existence. It empowers the conveyance of exact clinical data whenever any place by methods for cell phones. Ongoing innovative inventions in low-power integrated circuits, sensors, actuators, and remote communicability have increased the ease of designing a minimal effortless, compact, lightweight, and smart biosensor hub to a great extent. These hubs, equipped for detecting, preparing, and imparting multiple indispensable signs, can be flawlessly coordinated into remote individual or body zone systems for portable well-being checking.

In the paper Sannino and Pietro [8], a framework is recommended which is planned to screen patients' physiological boundaries to recognize strange cardiovascular increasing speeds and patient falls progressively. The center of this paper is the acknowledgment of a propelled versatile checking framework dependent on a measured programming engineering that rearranges mechanical or utilitarian changes. The paper additionally portrays a contextual investigation where the framework has uncovered significant advantages for both patients and clinical staff.

A vision-based fall detection system was proposed in the paper by Huang et al. [9] for patients or old people staying at home or at medicinal service establishments. This framework utilizes an omnidirected camera to ensure that every vulnerable angle gets covered. Additionally, this framework has considered any handy natural factors of the surroundings that may occur in a person's everyday life, for example, the event of light source flash, fluctuations in light that may affect the quality of camera's output. For better analysis, omnidirectional pictures are filtered and separated into radially and nonradially captured images directed as per the point related with a bodyline. Further, radially falling shapes are divided between internal and outward bearings. The identification highlights proposed for the framework incorporate variety in point and length related to the bodyline- and mobility-based pictures. Considering these highlights, an easy decision tree and custom threshold-based method is accepted for fall sensing. Test results prove that the proposed framework can take care of the issues of light source flicker and statically deserted articles.

The paper by Lv et al. [10] depicted the portable well-being observing device called iCare for the old. Remote body sensors and advanced mobile phones are tilized to screen the prosperity of the older. It can offer remote checking for the older whenever any place and offer custom-made types of assistance for every individual depending on their own well-being condition. When distinguishing a crisis, the advanced cell will naturally alarm related individuals who could be the elderly folk's individuals' loved ones and call the rescue vehicle of the crisis place. It additionally goes about as the individual well-being data framework and the clinical direction that offers one correspondence stage and the clinical information database so the loved ones of the served individuals can help out specialists to deal with him/her. The framework likewise includes some special capacities that take into account the living requests of the old, including customary update, fast alert, clinical direction, and so forth. iCare is not just ongoing well-being checking framework for the older, yet additionally, a living right hand that can make their carries on with increasingly helpful and agreeable.

Today, there is a requirement for a coordinated arrangement that can identify and analyze the early cyanotic innate illnesses in newborns to observing of different geriatric disorders in older individuals. Future approaches of higher life expectancy and uber-prompt therapy rely on it. The paper by Gahlot et al. [11] proposed a mechanism needed to make smart villages and cities in terms of the best healthcare sector. This system is proposed with appropriate observations and technological support of survey-based research expertise including questionnaires, assessments as well as systematic examination of problems in existing advancements.

IoT is increasing a quickly developing importance in numerous controls, particularly in customized medicinal services. In the meantime, body area sensor network (BASN), using IoT-based structure, has been broadly applied for omnipresent well-being analysis. Electrocardiogram (ECG) checking is accepted as indispensable methodology for the diagnosis of coronary illness. The principal commitment of the paper done by Wan et al. [12] incorporated the accompany: foremost, it proposes a novel framework, the wearable IoT-cloud-based health checking framework (WISE), ensuring an individual's good well-being status. It accepts the BASN structure for constantly analyzing the patient's well-being status. A few wearable sensors have been installed, which includes the HR, internal heat level, and the circulatory strain sensors. Besides, most of the existing wearable well-being observing frameworks need an advanced mobile phone for information handling, representation, and transmission door, which will surely affect the typical day-by-day utilization of the personal digital assistant (PDA), while in WISE, information gathered using the BASN is being continuously sent to the cloud, and a compact, light wearable liquid crystal display (LCD) to be implanted for easy access of data practically.

The paper by Dhakal et al. [13] intended to improve dependability and inactivity and to decrease power utilization by sensors during information's movement in WSBN (Wireless Body Sensor Network) by implementing enhanced reliability, energy-efficient, and latency (EREEAL) algorithm. Framework proposed in this paper comprises of an EREEAL's methodology to decrease gaps in information and start to finish delay with the enhancement of the dependability on remote communications in WBSN by transmitting the sensor information during various schedule openings utilizing time division multiple access (TDMA) examination and by limiting excess delicate information. The outcome proves that this new methodology enhances accuracy to 98% over the information bits produced. The proposed framework focuses on reducing impedance with information among sensors and centers around limiting information loss during the transmission.

The paper by Durán-Vega et al. [14] presented a proposition for real-time well-being checking arrangement for senior citizens living in geriatric homes. This framework was created to assist guardians with having a superior control in observing the well-being status of the patients and have nearer correspondence with their relatives of the patient. To approve the achievability and viability of the proposition, a model was fabricated, utilizing a biometric armband associated with a versatile application that permits the continuous representation of all the data produced by the sensors (pulse, internal heat level, and blood oxygenation) in the wristband. Utilizing this information, health workers can decide for the well-being condition of their patients. The assessment found that the clients saw the framework to be smooth and simple in terms of learning and utilizing, giving introductory proof that this proposal can enhance the nature of the grown-up's health.

The rising inescapable figuring is viewed as a promising answer for the frameworks of individual social insurance and crisis help, which can screen individual well-being status in a constant way and consequently issue the alarm for clinical guides if there should be an occurrence of a crisis. Be that as it may, the usage of such a framework is not trifling because of the issues, including (i) the framework activities without confused parental figure help to lessen the cost of the human asset, (ii) indispensable sensor organization for client comfort, (iii) on location fundamental information procedure and capacity for vitality sparing in information transmission, and (iv) opportune correspondence with medicinal service places for revealing clients' well-being status and looking for pressing clinical guides. In the paper by Wu et al. [15], a novel wearable individual medicinal service and crisis help framework was proposed, to be specific WAITER. This framework does not require explicit parental figure help. The server just utilizes the little wearable sensors to persistently gather the client's fundamental signs and Bluetooth gadget to transmit the tactile information to a cell phone, which can perform nearby crucial information stockpiling and procedure. After nearby information process, the cell phone can utilize its GSM module to intermittently report clients' well-being status to the medicinal service places and issue alerts for clinical guides once identifying crisis.

The paper by Alnosayan et al. [16] proposed the system named as MyHeart that refers to a remote health monitoring framework intended to connect the current hole in the congested heart failure care continuum that happens in the case of a patient advancing from the medical clinic to the home condition. This telehealth framework utilizes remote well-being gadgets and a versatile application on the patient's end, a standardized master framework, with a dashboard on the doctor's conclusion, thus supporting the to-and-fro transmission of health data relating to the side effects, vitals, well-being hazards, and consequences. This framework additionally gives messages to patients who wish to empower self-care according to Fogg's behavior model. Loma Linda University Medical Center is working on the verification of MyHeart's proposal and have also accounted for empowering beginning discoveries.

In the paper by Zhang [17], concentrating on the health analysis-related difficulties, another savvy computational methodology was proposed for efficient movement recognition, utilizing biomechanical elements upgrade and deep learning advancements. It can reveal profound concealed biomechanical designs from the cell phone-detected movement information and quickly identify 17 sorts of everyday and fall exercises performed by 30 individuals. The location precision of 11,770 exercises is as high as 93.9%, demonstrating the viability of the suggested idea. This exploration is relied upon to extraordinarily propel portable day-by-day movement and fall hazard observing in brilliant well-being time.

The review paper written by Appelboom et al. [18] discussed the various developments in monitoring systems and their utilities as smart wearable body sensors.

Health checking framework offers a ton of advantages to individuals' life particularly for the individuals who have an incessant ailment and need day-by-day perception. This well-being checking framework will improve personal satisfaction. Data innovation (IT) and remote correspondence framework can possibly improve security, quality, and productivity of medicinal services. By this improvement of the innovation, specialists and patients can without much of a stretch access and use well-being data varying. Binti Abu Bakar et al. explored [19] that well-being observing framework, specialists, and patients can utilize it so as to screen and check their condition through cell phone. The patients can be alarmed on the off chance that they are required to do the treatment or not. Arduino Uno and GSM along with temperature and HR sensors are utilized for this examination. Outcomes are right since the quantity of more seasoned individuals expands for a strange patient's condition.

Present day's IoT can be utilized to screen the well-being condition of the individual and monitor that to specialists or medical team remotely via IoT, as it is difficult to diagnose a person for 24 hours. So here the patient's well-being condition or health status, e.g., HR, respiratory rate, BT, body position, blood glucose, and ECG, can be estimated by using the nonobtrusive sensors. They are compatible with the Arduino's microcontroller board; it accumulates the data, for example, biomedical information using the sensors and the identified biomedical data can be sent to the server. Cloud platform, known as "ThingSpeak," was used in the paper by Kumar et al. [20] to upload the recognized data into the server. From this server, the data can be sent to the authorities and medical team through ThingSpeak's application. Along with this, using this smart well-being checking framework decreases the effort of experts and medical teams used in diagnosing their patients for 24 hours and moreover diminishing any delays, simultaneously making it cost-effective.

The purpose of the work which was done by Mahadewaswamy and Durgad [21] was to plan and build up a framework that plays out the errands of the estimation of imperative boundaries, for example, temperature and pulse. The deliberate qualities are constantly refreshed on ThingSpeak server by utilizing Wi-Fi module, and if there should be an occurrence of variations from the norm, the data are passed on to the associated specialist via SMS by utilizing GSM module. An alarm call is likewise started to the specialist following a few minutes. The framework additionally involves an inbuilt pill tracker unit that makes the patient at the perfect time aware of taking the endorsed tablet and it likewise cautions the patient if there should arise an occurrence of wrong tablet determination. All the while, the message with respect to the incorrect tablet choice is communicated to the overseer. This framework is likewise modified to quantify and arrange the idea of the pulse. The exhibition of the framework for the estimation of pulse and temperature is assessed under the management of a certified doctor. The readings measured using this framework are contrasted and the perceptions recorded by the doctor utilizing traditional strategies. The utilization of the framework as pill tracker is concentrated by thinking about a patient with different infirmities.

A signal-based secure connection framework was proposed in the paper by Rahman et al. [22] with shrewd home IoT well-being gadgets to help older individuals or individuals with exceptional requirements. The system utilizes a decentralized blockchain agreement for putting away the smart home IoT well-being information and client characters. The system influences off-chain answer for putting away crude mixed media IoT tangible payload and signal information. Utilizing the proposed well-being checking structure, a keen property holder or specialist organization can make a digital physical space with a protected advanced wallet for every human inhabitant and approved IoT well-being gadgets. Numerous approved home inhabitants can associate with the IoT-based keen home observing sensors, do client and IoT well-being tactile media enrollment, and move value-based qualities by means of secure tokens, just as crude IoT well-being information payload through motion.

The paper done by Mulla and Mote [23] presented a scalable alert-based health monitoring system using Raspberry Pi, sensors, and buzzer which can be controlled by a Bluetooth-based Android app.

The paper written by Mahajan and Birajdar [24] portrayed an IoT-based well-being observing framework for constant ailments. Enormous information investigation has become a fundamental component while managing IoT-based observation of incessant ailments and for this, a visualized minimal effort adaptable design of distributed computing can give valuable help with the upkeep of electronic well-being records, large information examination, and telemedicine considering not many of its focal points remembering for request asset designation, boundless capacity limit, reinforcement, and recuperation. The imagined framework comprises a body sensor to organize which gathers continuous well-being information from temperature, SpO_2, and accelerometer sensor. The particle photon fills in as a detecting hub that forms this genuine time information utilizing unassembled ARM processor and transmits it over to the cloud utilizing Wi-Fi. The information can be envisioned on the dashboard utilizing the Adafruit Web server. Here the cloud can fill in as a sink hub that handles information stockpiling and processing. The framework is exceptionally valuable for patients experiencing heart failure, osteoporosis, and asthmatic issue.

The research in the paper carried out by Sharma et al. [25] assisted with improving the extent of IoT in healthcare institutions with a various scope of upgrades. This study presented a textile-based wearable system technology, intelligent medical boxes, unobtrusive biosensors, and a cloud computing architectural framework among different innovations and progression that would move the medical sector to unmatched statures as far as effectiveness and patient's comfort are concerned. This paper suggested changing the business by constant trade of information to flawlessly and proactively offer forecast, conclusion, and cure. The structure this paper presented can be suitably named as the Internet of Medical Things (IoMT) that opens an entirely different road for the patient–healthcare supplier interface (PHI) and wearable health technology (WHT).

Intelligent well-being observing framework with wearable sensors, advanced cell (android based), and the keen Web server was proposed in a paper by Rahman et al. [26]. An on-request information securing is incorporated in the framework to examine the patient's well-being condition intuitively. The proposed framework lessens the ordinary test cost of neighborhood or remote zone patients and improves social insurance framework. The sensors assemble the patient's physiologic boundaries (e.g., pulse, pulse, breath rate, and internal heat level) intermittently and move the information to the advanced mobile phones by means of Bluetooth module. The patient information is then transferred to the server database utilizing an android application. The server screens the database brilliantly to discover variation from the norm in the patient's physiologic boundaries. In the event that the server discovers variation from the norm in the database, it sends a warning to the specialist android application to inform the state of the patient. In light of the condition, the specialist may get some information about the malady indication or recommend anything. Thus, it proposes an intermittent checking of the database versatile to the patient condition.

The IoT is gradually affecting on social insurance part that incorporates both the specialist and patient fronts. The IoT additionally changes quiet consideration at home. Specifically, the IoT has been comprehensively applied to couple different accessible clinical assets and give satisfactory, dependable, and shrewd human service utilities to the old patients with a constant ailment. Sensor innovation gives the following clinical gadgets and improves the generally speaking proficiency. By the utilization of remote patient checking, expenses can be decreased by diminishing the quantity of emergency clinic visits. The proposed work conducted by Bangera and Dhiman [27] showed the well-being observing an old individual dependent on IoT, by gathering pulse, temperature and circulatory strain esteems from sensors, handling the information through microcontrollers, and storing the information on the cloud. Then again, the specialist can get to the patient's pulse and temperature esteems through the Website page and the message is sent to the patient/guardian.

The paper done by Siam et al. [28] presented a protected IoT-based well-being checking framework that abbreviates the separation between a patient and the pertinent clinical association. Crucial signs caught from sensors are prepared and scrambled utilizing Advanced Encryption Standard (AES) calculation before sending to the cloud for capacity. A node microcontroller (MCU) is used to complete the handling and encryption capacities and for giving availability to the cover over Wi-Fi. Moreover, a clinical authority can imagine the private well-being information in real time simply subsequent to giving decoding certifications. In addition, the proposed framework gives an alarm by sending an e-mail to some patient's family members or planning master if fundamental signs are outside the typical rates. The proposed framework gives protection, security, and constant availability for private well-being information records.

Wireless sensor networks (WSNs) for medicinal services have developed extensively in these years. Remote innovation has been created and utilized broadly for various clinical fields. This innovation gives human service administrations to patients, particularly who experience the ill effects of constant maladies. Administrations, for example, providing food, persistent clinical observing, and dispose of unsettling influence brought about by the sensors and instrumentation. Sensors are associated with a patient by wires and become bedbound that reduces the versatility of the patient. In the paper published by Ali et al. [29], continuous heartbeat checking framework was proposed by means of leading an electrical circuit design to gauge heart pulse (HP) for patients and show heartbeat estimating through cell phone and PC over the system progressively settings. In HP estimating application, sensor innovation is utilized to watch heartbeat by carrying the unique finger impression to the sensor by means of an Arduino based microcontroller with Ethernet shield which can interface heartbeat circuit to the Web and upload results to the Webserver and retrieve results it any place. This proposed framework gives the ease of use to a client (making it user-friendly). Likewise, it offers high speed and great precision, the most elevated accessibility with the client on a continuous premise.

Convenient savvy well-being observing framework with remote access to the doctor is extremely basic for elder residents and rustic human service staff. This can be accomplished by a low-power, precise, conservative, practical, easy-to-understand framework that is equipped for estimating the patient's fundamental boundaries. This information is shown locally at the patient's end and furthermore consistently sent to the doctors' end utilizing the IoT stage. The work done by Samik et al. proposed [30] that the framework utilizes noninvasive sensors to screen the imperative boundaries like pulse, SpO$_2$, and internal heat level precisely. "Raspberry Pi 3B+" minicomputer is utilized to collect, coordinate, analyze the sensors' information, and remotely impart to the clinical institutions or the doctor utilizing "MathWorks" cloud or applet. The dependability of the transferred essential parameters is additionally confirmed by the privately shown information. The utilization of Webhook applet to inform the doctor about the well-being state of the patient makes the suggested framework novel.

14.3 NOVELTY OF THE IDEA

As the title suggests, the primary objective of this device is concerned itself with the development of ambient-assisted healthcare monitoring system with IOT. The developing system can successfully detect and send real-time heath data to patient's relatives and doctors. Also, generating alarms in case of any health abnormality or emergency, which will allow the timely delivery of medical assistance, mitigates the long-term effects of these circumstances.

In this modern era of information and technology, we have seen a lot of advancements in the healthcare industry. Many gadgets and devices are being introduced to increase a patient's reach and control over his/her health. This section highlights the novelty and usefulness of this gadget.

14.3.1 OVERVIEW

In this project, an IoT-based smart health monitoring system for old people is developed. It first measures a patient's HR and BT and then detects free fall after analyzing the received data. Finally, it sends an alert via e-mail and SMS when those readings exceed foreordained critical values. An open-source IoT platform, ThingSpeak, is used for storing and retrieving the pulse rate and BT data. Therefore, patient's health can be observed and analyzed from any part of this world remotely via Internet. A panic button is also connected for the patients to press in case of an emergency which informs their relatives by sending e-mail/SMS alert. In case of a free fall, if a person fells, then alarm will be turned on for some seconds but if it was a minor fall and the person is fine, then he or she can switch off the alarm else the e-mail will be sent.

14.3.2 MULTIFUNCTIONAL

To solve the problem of elderly healthcare, two wearables connected by Bluetooth (for hand and waist) are designed. Thus, this gadget is multifunctional, which provides many features in a single device:

1. Fall Detection: This is done using accelerometer (ADXL145) to detect fall based on the changes in g-force values along x-, y-, and z-axes and send the message to cloud (which further sends alert to patient's relatives and doctors, if a fall is detected). Raw values of the sensors are sampled and calibrated to get more accurate data regarding the orientation of the object.

2. Body Temperature: LM35 is used to record the real-time temperature that allows the doctor to analyze the effectiveness of an ongoing treatment and detect an occurrence of a new illness. This is very useful, as studies have shown that unintended BT is a symptom for many diseases.

3. Pulse Detection: Pulse rate sensor tells the current pulse rate that helps give information about the blood pumping efficiency of heart and general health fitness level. Pulse rate is a significant factor representing a person's overall health.

4. Push Button (Panic): If the patient feels unwell/uncomfortable, thus wishes to see a doctor, then he or she can press the push button for panic. It will alert the health officials immediately at a high priority.

14.3.3 FEATURES

The following features distinguish this gadget from other already existing gadgets in the market. Thus, it highlights the idea's superiority over others:

1. Real-time health monitoring: Processed data of BT and pulse detection are sent continuously to health personnel and relatives of the patient. Thus, doctors can easily monitor patient's health in real time.

2. Easily interpretable: The graphical representation of health data makes it effortless for doctors to track and comprehend health data of many patients altogether easily and quickly.

3. Sending alerts and eliminating false positives: This gadget not only monitors real-time health but also sends alert in case of an emergency to doctors or family members. Alert is sent in the following cases:
   a. if the patient feels unwell and presses panic button.
   b. if BT or pulse rate goes beyond the preset critical values.
   c. if fall is detected by the sensor.
   In the latter case, small gap of a specific time frame is kept between the detection of fall and sending of alert. If it is not an actual fall (just a jerk/ body bent or an invalid fall detection), then the patient can press the button and alert will not be sent to health officials, thus avoiding any false positives or false alerts.

4. Simplistic and user-friendly gadget: Even after being multifunctional, it is not difficult to operate the gadget. It only has one push button for panic, thus increasing the comfort of our target customers that are senior citizens.

5. Cost friendly: This is a very important factor in any gadget's designing because increase in the product's price limits the customer. This is made cost friendly without compromising the quality and efficiency by making use of individual sensors with Arduino board.

14.3.4 Technology Stack

1. ThingSpeak IoT platform: This platform provides doctor and family members with the ability to view and analyze patient's health from any place using the Internet. All the sensor's data values are sent as parameters to this platform. To ensure privacy and security of user's personal data, this IoT analytics platform is secured by 16-digit Read API key.
2. Bluetooth: This technology is helping the two Arduino boards in communicating with each other. As this involves the development of two wearables, one for waist and other one for hand, a fast and secure medium of communication is maintained between them for sharing health data and sending it to cloud altogether.

14.4 DESIGN APPROACH

14.4.1 Components Used

1. Arduino Uno: Arduino Uno (based on ATmega328P) is an all-in-one-type integrated microcontroller board used for programming different sensor-based circuits and many other purposes. It consists of 14 digital pins for input/output (six of them can be used for pulse-width modulation (PWM)-based outputs), six analog inputs, a 16 MHz quartz crystal, a USB connecter, a power jack, a reset button, and an in-circuit serial programming (ICSP) header.
2. ESP8266: ESP8266, also known as Wi-Fi module, is a well-designed microcontroller, which is developed by a China-based company, Espressif Systems, as shown in Figure 14.2. This low-cost Wi-Fi microchip has the ability to connect to a Wi-Fi network, and thus, it makes TCP/IP connections more easy and simple.
3. Pulse Rate Sensor: This is a well-known plug-and-play-type sensor for measuring pulse rate to be used with a microcontroller as shown in Figure 14.3. The sensor can be easily clipped onto a fingertip or earlobe and plugs right into Arduino. Front side of the sensor features a heart-shaped logo which makes contact with the patient's skin. Front side also has small round hole from which the back LED shines and below which lies the ambient light sensor similar to the one found in our mobile phones, laptops, and tablets in order to control the screen's brightness under different light conditions. Rest of the parts are mounted on the other side of the screen. When LED glows, its light falls onto the earlobe/fingertip or capillary tissues, then it detects the amount of light that is reflected back in order to calculate the noninvasive HR. An open-source monitoring app can be used for plotting the patient's pulse in real-time T as graphs help in clear and better understanding of the numerical data.
4. LM35 Temperature Sensor: It is a high-precision temperature sensor. It gives analog voltage as output which is directly proportional to the temperature (in the unit of degree Celsius). Its temperature range for operation lies from −55°C to 150°C. Its stand-by current is less than 6 µA and can be operated from either a 5 V or a 3 V power supply. For every 1°C rise or fall in temperature, its output voltage varies by 10 mV as shown in Figure 14.4.

FIGURE 14.2 W-Fi module (ESP8266).

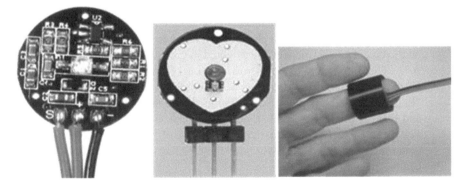

FIGURE 14.3 Pulse rate sensor.

14.4.2 METHODOLOGY

ThingSpeak is a well-known platform service for IOT-based projects. It has integrated support from MATLAB, popular numerical computing software. Data of this health monitoring system can be remotely monitored and controlled via Internet, by making use of ThinkSpeak's Webpages and channels. ThingSpeak analyzes and visualizes the information after collecting it from different sensors and finally gives the output by triggering an alert-based reaction. It helps in the creation of different location-tracking applications and sensor-logging applications, with real-time updating of data.

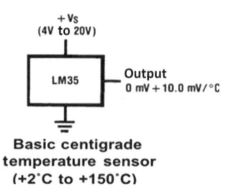

Basic centigrade temperature sensor (+2°C to +150°C)

FIGURE 14.4   Temperature sensor.

Here, ThingSpeak platform is utilized to monitor the patient's health by recording and analyzing the pulse rate and BT by making use of the Internet. IFTTT, a freeware Web-based service, will be used in interfacing ThingSpeak software to the associated message/e-mail service with the goal of sending an alert whenever the patient needs diagnosis like if a fall is detected or he or she is found in some critical state.

Configuration of ThingSpeak to record and analyze the patient's data on the Web:

Step 1: - Firstly, the client needs to sign up by creating a fresh account on the portal of ThingSpeak.com, then tap on the green-colored box of "Get Started" after logging in to that account.

Step 2: - After signing in, client should open the menu of "Channels" and tap on the box of "New Channel" highlighted in green color.

Step 3: - This step will take the user to a channel creation form, complete the fields of Name and Description as per individual's choice. And then, enter "Pulse Rate," "Temperature," and "Panic" in the labels of Field 1, Field 2, and Field 3. Subsequently, click on the checkboxes of these three fields to tick them. Likewise, checkmark the check box of "Make Public" provided underneath the form and lastly Save the Channel so that, a fresh channel can be created.

Step 4: - Now, three empty charts can be seen. Take note of the Write API key, as this key will be utilized for coding.

Step 5: - Here, ThingHTTP server application will be utilized to launch the IFTTT applet for collecting information to Google Sheets and sending of SMS/e-mail.

Figure 14.5 shows applets created in IFTTT server. ThingHTTP executes interdevice communications among different gadgets, devices, Web-portals or Websites, and Web services without the need of executing that protocol on device level. One can choose or specify activities in ThingHTTP, which he or she wants to implement utilizing other ThingSpeak applications, like React. In order to create "New ThingHTTP," required URL for triggering can be collected from IFTTT.

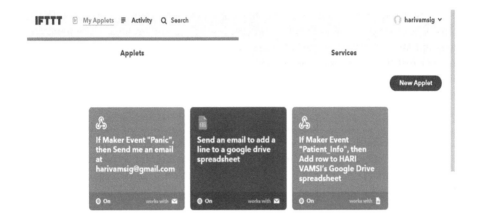

FIGURE 14.5 Applets created in IFTTT server for creating Google Sheets and sending mail.

Configuration of IFTTT portal for activating e-mail service using ThingSpeak Values:

Step 1: - Log onto the IFTTT, find "Webhooks" using search box and open it.

Step 2: - Tap on the Documentation.

Step 3: - In the event box, type "Patient_Info" and copy-paste the shown URL to some different locations as it will be utilized later for ThingHTTP.

Step 4: - Choose the option of "New Applet" under My Applet section.

Step 5: - Tap on "+this" then, look for Webhooks and open it. Pick trigger of "Receive a web request."

Step 6: - Tap on "+that," then find Google Sheets, and open it. After opening, insert row into the spreadsheet.

Step 7: - Type a suitable name for the Google Sheets. Under the label of "Formatted Row" box, date and time of the occurrence, event occurred, value of pulse rate, and BT will be seen.

Step 8: - Finally finish the process after reviewing.

14.4.3 IMPLEMENTATION

Now that the root issue has been built up, a technique for detecting/sensing fall is distinguished. The significant priorities for identification of falls are as follows:

   i. Nonobtrusive and intuitive

  ii. Wearable framework

 iii. No false positives

 iv. No defer time

  v. User-friendly by involving a common method of communication.

In the realm of innovation, wearable has consistently had an exceptional intrigue. This is because of the simple entry and use, just as the unified idea of the item. One does not endure the worst part of social disgrace by wearing such a gadget, in this manner fulfilling the client requirement. Principal task is the accuracy and efficient productivity of the device, as the message must be conveyed from direct A toward point B in a snap of a finger. A setup method of communication must be utilized; for this situation, the e-mail can contact individuals the speediest.

Bogus positives can be both awfully irritating and burning through both time and cash. So, in request to guarantee precision, the fall-identification mechanism must be idealized, and the gadget must be made user-friendly and intuitive. Taking all the above elements into thought, the idea of utilization of an accelerometer-based detecting gadget around the waistline is suggested. This would not just copy the appearance of an ordinary belt, but at the same time it is the area where focal point of gravity lies; therefore, readings that one gets from the sensor will be exact and trustable. Database is saved at the doctors' end, which gives a constant stream of data associated with patient's health. In case of any issues detected, a suitable alert is issued both at the central display, as well as via e-mail to the doctors and relatives. E-mail is the most rudimentary, ever-accessible means of communication in the modern world and cannot be missed by any individual. IoT is used to communicate from machine-to-machine. Various available modules can be employed such as Wi-Fi-modules (e.g., ESP8266), zigbee protocol, and Ethernet shields. However, for the purpose of prototyping, a Bluetooth serial transfer protocol between the microcontrollers is implemented.

14.5 CONCLUSION

14.5.1 Results

As soon as the hardware connections are done. Readings for change in position, temperature, and pulse rate can be seen in serial monitor of Arduino.

1. If any free fall is detected, a fall line can be seen in the serial panel. Consequently, an e-mail will be sent to hospital and the patient's relatives stating the condition of the patient, according to which required actions can be taken. Figure 14.6 shows Arduino's readings inside the serial monitor.
2. As a result of continuous monitoring, whenever the value of temperature goes above the threshold value, an e-mail will be sent to the hospital and the patient's relatives stating the condition of the patient. Similarly, when the pulse value goes above the threshold value, an e-mail will be sent to the hospital and patient's relatives stating the condition of the patient, so that required actions can be taken. Figure 14.7 shows a sample mail which is sent when patient presses the panic button seeking for help.
3. Figure 14.8 shows data of pulse rate sensor, temperature sensor, and panic button in an understandable and readable format. Continuous monitoring is supported by ThingSpeak cloud services, which can be directly analyzed by the medical team.

FIGURE 14.6 Arduino's serial monitor showing readings.

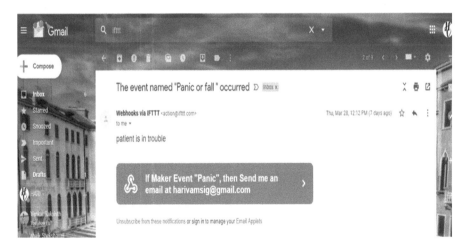

FIGURE 14.7 Alert mail sent in case of fall or panic.

4. Thus, fall detection circuit attached with the belt combined with temperature sensor and pulse sensor attached to the wristband together makes smart health wearable for elderly people as show in Figure 14.9.

14.5.2 Summary

In this study, outlooks of the project using the concept of IoT to develop a healthcare monitoring system for the elder citizens is presented. Therefore, the key goal of this project was to develop an efficient healthcare system for the elderly with concomitant ease of use. As shown in Table 14.2, designed system's cost can be approximated

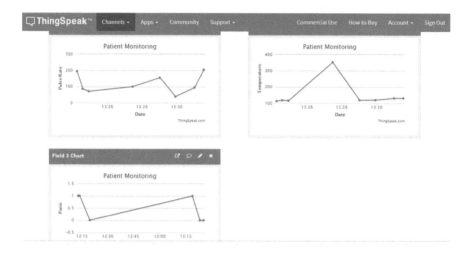

FIGURE 14.8   Graphical representation of pulse rate, temperature and panic data.

FIGURE 14.9   Wristband and waistband connected using Bluetooth technology.

after performing cost analysis. This system can effectively detect and produce warnings in case of any health abnormality or emergency, that will opportune the timely delivery of medical assistance and help diminishing the chronic consequences of these situations. The IoT enables us to collect and share the information retrieved from the wearable healthcare devices.

TABLE 14.2
An Approximation of Designed System's Cost Analysis

Items	Quantity	Price	Total
Arduino Uno	2	510	1020
EP8266 Wi-Fi module	1	275	550
LM35 temperature sensor	1	70	70
Pulse rate sensor	1	200	200
Push button	1	5	5
Bluetooth	2	200	400
Accelerometer	1	150	150
Connecting wires	2 sets	20	40
Breadboard	1	60	60
Total			2,495

These data are then shared with the patient and their respective medical personnel. The chain of notifying the time to time updates make the communication faster, feasible and inhibits any sort of inaccuracy in the state of emergency. IoT has a major role to play in the field of healthcare and medical recovery. IoT uses intelligent remote technologies that can enhance early diagnostics, accuracy in providing data and immediate care better life.

REFERENCES

[1] M. M. Islam, A. Rahaman, M. R. Islam (2020). Development of smart healthcare monitoring system in IoT environment. *SN Comput. Sci.* 1, 185.
[2] M. Al-khafajiy, T. Baker, C. Chalmers, M. Asim, H. Kolivand, M. Fahim, A. Waraich (2019). Remote health monitoring of elderly through wearable sensors. *Multimed. Tools Appl.* 78, 24681–24706.
[3] T. Tsukiyama (2015). In-home health monitoring system for solitary elderly. *Procedia Comput. Sci.* 63, 229–235.
[4] A. Bourouis, M. Feham, A. Bouchachia (2011). Ubiquitous mobile health monitoring system for elderly (UMHMSE). *Int. J. Comput. Sci. Inf. Technol. (IJCSIT)* 3, 3.
[5] S. S. Bhadoria, H. Gupta (2013). A wearable personal healthcare and emergency information based on mobile application. *Int. J. Sci. Res. Comput. Sci. Eng.* 1, 4, 24–30.
[6] S. J. Park, M. Subramaniyam, S. E. Kim, S. Hong, J. H. Lee, C. M. Jo, Y. Seo (2017). Development of the Elderly Healthcare Monitoring System with IoT. In *Supporting Phobia Treatment with Virtual Reality: Systematic Desensitization Using Oculus Rif*, pp. 309–315.
[7] R. Shahriyar, M. F. Bari, G. Kundu, S. I. Ahamed, M. M. Akbar (2009). Intelligent mobile health monitoring system (IMHMS). *Electronic Healthcare: Second International ICST Conference, eHealth 2009*, Istanbul, Turkey, Revised Selected Papers (pp. 5–12).
[8] G. Sannino, G. D. Pietro (2011). Advanced mobile system for indoor patients monitoring. *2nd International Conference on Networking and Information Technology (ICNIT 2011)*.
[9] Y.-C. Huang, S.-G. Miaou, T.-Y. Liao (2009). A human fall detection system using an omni-directional camera in practical environments for health care applications. *MVA2009 IAPR Conference on Machine Vision Applications*, Yokohama, Japan.

[10] Z. Lv, F. Xia, G. Wu, L. Yao, Z. Chen (2010). iCare: a mobile health monitoring system for the elderly. *2010 IEEE/ACM International Conference on Green Computing and Communications & 2010 IEEE/ACM International Conference on Cyber, Physical and Social Computing.*

[11] S. Gahlot, S. R. N. Reddy, D. Kumar (2019). Review of smart health monitoring approaches with survey analysis and proposed framework. *IEEE Internet Things J.* 6, 2, 2116–2127.

[12] J. Wan, M. A. A. H. Al-awlaqi, M. Li, M. O'Grady, X. Gu, J. Wang, N. Cao (2018). Wearable IoT enabled real-time health monitoring system. *J. Wireless Comput. Network* 2018, 298.

[13] K. Dhakal, A. Alsadoon, P. W. C. Prasad, R. S. Ali, L. Pham, A. Elchouemi (2020). A novel solution for a wireless body sensor network: telehealth elderly people monitoring. *Egyptian Inf. J.* 21, 2, 91–103.

[14] L. A. Durán-Vega, P. C. Santana-Mancilla, R. Buenrostro Mariscal, J. Contreras-Castillo, L. E. Anido-Rifón, M. A. GarcíaRuiz, O. A. Montesinos-López, F. Estrada-González (2019). An IoT system for remote health monitoring in elderly adults through a wearable device and mobile application. *Geriatrics* 4, 14.

[15] W. Wu, J. Cao, Y. Zheng, Y.-P. Zheng (2013). WAITER: a wearable personal healthcare and emergency aid system. *Int. J. Sci. Res. Comput. Sci. Eng.* 1, 4, 2320–7639. 1, 4, 23207639.

[16] N. Alnosayan, E. Lee, A. Alluhaidan, S. Chatterjee (2014). MyHeart: An intelligent mHealth home monitoring system supporting heart failure self-care. *2014 IEEE 16th International Conference on e-Health Networking, Applications and Services,* Healthcom.

[17] Q. Zhang (2019). Deep learning of biomechanical dynamics in mobile daily activity and fall risk monitoring. *IEEE Healthcare Innovations and Point of Care Technologies, (HI-POCT),* Bethesda, MD, pp. 21–24.

[18] G. Appelboom et al., (2014). Smart wearable body sensors for patient self-assessment and monitoring. *Archives of Public Health* 72, 28.

[19] N. H. Binti Abu Bakar, K. Abdullah, M. R. Islam (2016). Wireless smart health monitoring system via mobile phone, *2016 International Conference on Computer and Communication Engineering (ICCCE),* Kuala Lumpur, pp. 213–218.

[20] S. P. Kumar, V. R. R. Samson, U. B. Sai, P. L. S. D. M. Rao, K. K. Eswar (2017). Smart health monitoring system of patient through IoT, *2017 International Conference on I-SMAC (IoT in Social, Mobile, Analytics and Cloud) (I-SMAC),* Palladam, pp. 551–556.

[21] U. B. Mahadewaswamy, Z. Durgad (2018). IoT based smart health monitoring system. *i-Manager's J. Electron. Eng.* 9, 2, 8.

[22] M. A. Rahman, K. Abualsaud, S. Barnes, M. Rashid, S. M. Abdullah (2020). A natural user interface and blockchain-based in-home smart health monitoring system, *2020 IEEE International Conference on Informatics, IoT, and Enabling Technologies (ICIoT),* Doha, Qatar, pp. 262–266.

[23] A. Mulla, P. T. Mote (2016). Smart (scalable medical alert response technique) Health Monitoring System Using Raspberry Pi, pp. 490–494.

[24] S. Mahajan, A. M. Birajdar (2019). IOT based smart health monitoring system for chronic diseases, *2019 IEEE Pune Section International Conference (PuneCon),* Pune, India, pp. 1–5.

[25] A. Sharma, T. Choudhury, P. Kumar (2018). Health Monitoring & Management using IoT devices in a Cloud Based Framework, *2018 International Conference on Advances in Computing and Communication Engineering (ICACCE),* Paris, pp. 219–224.

[26] Z. Rahman, I. Husain, E. Haque, F. Enam, A. Wadood (2019). Smart Health Monitoring with On-demand Data Acquisition and Analysis.

[27] V. G. Bangera, V. Dhiman (2017). IOT based smart health monitoring for old aged patients, *Int. J. Recent Trends Eng. Res.* 3, 5, 431–437.

[28] A. I. Siam, A. Abou Elazm, N. A. El-Bahnasawy, G. El Banby, F. E. Abd El-Samie, F. E. Abd El-Samie (2019). Smart health monitoring system based on IoT and cloud computing, *Menoufia J. Electron. Eng. Res.* 28, 37–42.

[29] N. S. Ali, Z. A. Alkaream Alyasseri, A. Abdulmohson (2018). Real-time heart pulse monitoring technique using wireless sensor network and mobile application, *Int. J. Electr. Comput. Eng. (2088-8708)* 8, 6.

[30] B. Samik, M. Ghosh, S. Barman (2020). Raspberry PI 3B+ based smart remote health monitoring system using IoT platform. *Proceedings of the 2nd International Conference on Communication, Devices and Computing*, Springer, Singapore, pp. 473–484.

15 Energy Management System for Smart Buildings
Mini Review

Shivangi Shukla, V. Jayashree Nivedhitha,
Akshitha Shankar, and O.V. Gnana Swathika
Vellore Institute of Technology

CONTENTS

15.1 INTRODUCTION

Energy management is the need of the hour. With the increase in demand and consumption with advancements in technology, efficient management of the existing amount of energy has become a necessity in the modern times. Smart buildings require different amounts of energies at different but continuous time frames from varied sources in order to sustain the functioning of their smart devices and to ensure smooth load scheduling operations.

Among all the fields that require energy, buildings and localities require a majority of the generated quantity. According to estimates, buildings are accountable for around one-third of the total energy consumption in the USA. A microgrid management system is a residential building that has a two-way communication architecture with different load and supply sources [1].

However, the generating authority must be responsible for many factors such as the price, rate of consumption, limited uptime and downtime, with varying gradient of the energy demand profiles. Additionally, there are also a lot of stability and reliability issues with the smart grid. Thus, it becomes very much vital to incorporate optimization techniques, methodologies, and smart technologies, which are to be answerable to the demand and to sustain the generation without constraints [2].

With the increase in the availability and development of renewable energy sources, energy management systems (EMSs) in smart buildings are being designed and implemented with better and improved results as the days pass by with the advent

DOI: 10.1201/9781003240853-15

of smart technologies, algorithms, and reliable methodologies. The main aim of all these systems is to reduce the cost of electricity consumption and to improve the load characteristics without compromising the end user comforts. For example, the main ideology as described by [3] is to design an EMS that reduces the peak demand as well as the electricity price by following source and load scheduling control algorithm. The source scheduling algorithm works in a way to charge the battery bank (with hybrid car) during low price electricity hours and acts as a source during high price electricity hours. The load scheduling algorithm on the other hand shifts the suitable load from peak hours to regular hours, thereby reducing demand, price, as well as gap in energy exchange utility band using renewable power supply like PV as primary sources, for DC microgrid loads.

Many techniques and methodologies are used in recent times to efficiently manage energy in smart buildings. The techniques proposed in [4] provide for utilizing network nodes in a BAS such that low cost is provided. Nodes can be used to determine the location of other nodes within the same block. Information can also be shared between the other nodes during commissioning the process. In addition to this, ranging techniques can be used to track nodes.

Smart buildings comprise not only residential buildings but also other prime buildings such as hospitals and educational institutions. Reference [5] highlights the idea of an intelligent EMS to harness the benefits of all subsystems in a hospital, and facilitating effective operation, while also ensuring that the operations of the hospital are not affected during extreme conditions such as natural disaster. There is a need to achieve the best healthcare services and, at the same time, reduce their costs without affecting the quality. This enhances flexibility and increases environmental sustainability. This is a step toward zero-energy buildings, shifting the focus from diesel-based systems that are still being used in most cases.

High Voltage Alternating Current (HVAC) equipment consumes a lot of energy, and smart buildings account for a large share in the total energy consumption. With the wide range of energy sources and advancements in technological and social fronts, it has become a necessity to bring in new EMSs such as distributed energy sources or hybrid energy sources. Smart grids are an important advancement in terms of energy management [6].

Smart buildings are a major part of the smart grid architecture. Thus, proper management of these grids is also of utmost importance. They consist of smart buildings, electric vehicles, residential sectors, and localities consuming energy along with the incorporation of information and communications technology (ICT). Many algorithms and methods have been proposed and implemented to ensure higher grid stability and also to reduce the increased consumption of fuel.

For example, in [7], the authors concentrate on the design of a two-level EMS for a DC microgrid that overcomes the difficulties of meeting the energy requirements of a multi-source hybrid DC microgrid in the traditional distributed control method.

Greenhouse gas (GHG) emissions have led to the increase in the use of electric vehicles. But the improper schedule of these vehicles has led to grid failure in many cases. Thus, to effectively reduce the chances of microgrid failures to which these vehicles are connected, many methodologies such as the introduction of xEVs and renewable power supply sources have been discussed and implemented in [8].

Apart from the usual smart grids, there are islanded grids that are disconnected from the main grid and operate independently with power sources and load. Many EMSs also concentrate on increasing their efficiency. In [9], the authors propose a method to satisfy the need for islanded microgrids that are used as flexible, adaptive, and sustainable smart cells of distribution power systems, operated for technical and economic purposes. Besides, a new demand response, dependent on frequency response, is conducted to the EMS to cope economically with uncertainty. This co-optimizes the microgrid energy resources.

EMS has been in existence for a long time, but there is a need to improve its design and output characteristics to improve stability and reliability.

The primary functions of smart home energy management system (SHEMS) are to monitor, manage, and improve the flow and use of energy. The SHEMS consists of many components such as measuring device, sensing device, ICT, smart devices, and EMS [10].

Thus, it is of utmost importance to develop advanced EMS using optimization algorithms, smart technologies, and other methodologies as discussed in this paper. All these ideas and implementations aim to decrease cost and increase comfort also has a constant check on the environment.

15.2 METHODOLOGIES

Many researchers and field experts have proposed and analyzed various methods that can be incorporated in energy management for smart buildings. All these methods have proved to be cost-efficient and have given desirable results in terms of energy efficiency and management. References [11,12] use mixed integer linear programming (MILP) to develop the proposed model.

Reference [11] enlightens about a two-way framework in energy management of a university campus connected to a smart grid using renewable energy sources, storage system and EVs used for transportation, using MILP framework that controls the power flow in the grid. Different schedules of the EVs have been planned in the framework in order to efficiently use energy. Multiple scenarios based on the availability of the renewable energy sources are also considered using previous year's PV data to design the system where cost reduction has been immensely observed and good amount of energy transfer back to the grid has also proved to be efficient, whereas in [12], the main aim of the system was to reduce the consumption cost and also to predict unexpected outrage in the microgrids (security-based equations) by using an algorithm named MILP in modeling for proper energy management (optimally) of residential sector microgrids based on unbalanced three-phase distribution systems with single-phase PV resources and single-phase energy storage sources. The model was developed after linearizing the nonlinear model.

In [13], a smart building cluster is described in order to effectively reduce power consumption and to improve load characteristics. All the smart buildings are considered as players in the non-cooperative game theory with integrated PV system and automated demand response (ADR). Nash equilibrium model is proved by multi-objective optimization problem (MOP) theory, which considers many factors such as electricity price, cost model, subsidy, and user convenience, which are implemented on a real model. This multi-party arrangement witnessed decrease in consumption as well as improvement in load factor.

The plan mentioned in [14,15] concentrates on AC/DC grid management. The idea in [14] is mainly aimed for rural areas, which shall be connected to the grid which explains the design and purpose of AC/DC grid EMS that reduces the power exchanged between grids by optimal use of renewable resources (photovoltaic and wind) by maintaining state of charge (SOC) in battery banks and fuel cell as a complimentary which is implemented using multiple particle swarm algorithm that ensures maximum utilization of resources and the desired results are also observed for 15 different modes of operation. On the other hand, in [15], the authors talk about a complex grid structure where maintaining the system stability and safety is of utmost importance. They have designed a hybrid AC/DC architecture of a microgrid suited for smart buildings so that grid isolation can be maintained, grid interference can be avoided, and the penetration of distribution generators can be improved. Therefore, the safety of the system and stability can be confirmed in a complicated grid environment. The characteristics of operating modes along with the comprehensive control strategy and the attribute of every single mode are designed and validated using a simulation platform.

In [16], the authors proposed an idea of a smart home model including many energy resources and practical comfort specifications that the user can freely use. An automated home energy management system (AHEMS), with properly suited algorithms, must permit for integrated optimization of several energy resources such as grids and storage, in face of modified tariffs. An AHEMS minimizes the total cost and dissatisfaction caused to the users and provides compromising solutions by rescheduling load operations, etc. The result analysis shows that with the deployment of AHEMS, unbalanced scenarios due to fixed power supply can be balanced and safely substituted by considering scenarios of variable power charge. This also shows that the computing time for this AHEMS is 90s.

Reference [17] describes a plan that aims to develop a hybrid EMS architecture based on canonical coalition games. A central EMS along with a local EMS performs scheduling, monitoring, and rescheduling functions and complements the supply process. Whenever a disruption is identified, the EMS mentions its power surge or its deficit for every time duration of the rescheduling period and this information is updated on a constant basis. Cooperative management is based on canonical coalition games, which ensures that the best coalition structure is the grand coalition structure [17].

Reference [18] explains the efficiency of a smart EMS of a DC microgrid that can divide the demand across many generators. The EMS is developed for energy monitoring and voltage control and also to produce a balance between demand and supply using a reconfiguration hierarchical algorithm. A controller area network (CAN) bus is used to deliver communication between them. Distributed energy resources that primarily contribute to the demand in microgrids in green buildings are linked together by power electronic converters, which are controlled by smart controllers. The stability and robustness of the control methods determine the efficiency of this converter, and thus, this framework offers online control of microgrids, simple implementation, high reliability, and ease of redevelopment as advantages.

In [19], the methodology is to combine smart thermostat with home energy management system (HEMS), whose joint usage provides a greater benefit. It combines day-ahead load scheduling for reduction of cost, and the fuzzy logic-based thermostat aims for thermal comfort. These happen in two stages where different set points are defined per time interval by the thermostat, by fuzzifying input parameters and occupant presence that is used in the second stage for scheduling purposes. Effect and validation of the planned smart HEMS thermostat are assessed under varying solar radiations and temperature, and a simulation of this model shows a significant cost reduction compared to other conventional thermostats.

A user-centric smart solution as proposed in [20] assists energy sustainability of modern cities. Here, in a smart building, the building management platform has been deployed in which a set of few tests were conducted. The concerns in the building's infrastructure management were assessed, by carrying out tests. The first stages showed an energy saving in heating of around 20% at building level, which translates into an 8% reduction in the building's energy usage.

As per the proposed method in [21], the different blocks of buildings need to assure energy efficiency and sustainability since the power system undergoes transition from a "centralized" + "vertical" architecture to a "decentralized" + "horizontal" one. Building energy management system (BEMS) shows guaranteed flexibility for demand-side management and demand responses (DSM&DR) in the built environment, while interacting with smart grids. BEMS is flexible to be optimized to capitalize the built environment's demand in the smart grid context, using some control techniques while ensuring that there are minimum comfort levels required by the consumers. Here, a control strategy that is agent-based is proposed for the functioning of buildings. Thus, a decision-making strategy based on fuzzy rule is made for monitoring and controlling the energy flows in a building [21].

A hybrid intelligent system that uses a fog-based architecture to accomplish energy efficiency in smart buildings is proposed by authors in [22]. It combines reactive intelligence and deliberative intelligence for faster adaptation to constantly changing environment and for performing very tough learning and optimization, respectively. This hybrid characteristic permits the system to adapt well by reacting in real time to necessary scenarios and to constantly better its performance by learning upon the user's needs. The effectiveness of this approach is verified by extensive experiments on real sensor traces in the smart homes. Satisfying users' needs and preferences, experimental results prove that this system achieves economical energy savings in its smart environment management.

The Goldwind microgrid testbed uses energy management architecture. In [23], a complete mathematical model is described by the authors, including all constraints, for microgrid operational management. For each span of 10 minutes, prediction results are obtained using the modified fuzzy prediction interval model. Particle swarm optimization algorithm is used to train this model. The maximum transaction cost scenario provides the worst-case operating points. Considering a lot of uncertainties in load demand and renewable power generation, this operating point (maximum transaction) shields those factors more efficiently compared to other scenarios.

A price-sensitive operational model is developed in [24], for HVAC systems of building while at the same time considering rigid loads and distributed power resources like battery storage for the buildings and PV generation. To minimize the net cost of energy usage, a nonlinear economic model predictive controller is developed by HVAC system of building without compromising on the comforts of users. The efficiency of this controller is verified using case studies simulations.

15.3 OPTIMIZATION AND ALGORITHMS

An optimized home energy management system (OHEMS) that aims in incorporating the residential arena into demand-side management activities by the combination of renewable energy sources and energy storage systems is developed in [25]. This system reduces the electricity bill considerably by prioritizing and scheduling the smart building's devices along with ESS in accordance with the changing price of the electricity rates. This optimization problem is designed by various knapsack problems and solved by using the heuristic algorithms, genetic algorithm (GA), wind-driven optimization (WDO), bacterial foraging optimization (BFO), binary particle swarm optimization (BPSO), and hybrid GA-PSO (HGPO) algorithms.

The modeling of energy management in smart buildings consisting of renewable photovoltaic resources and responsive/non-responsive devices is addressed in [26]. Employment of the solar power system is managed using the KNX protocol. The primary objective is to reduce the losses incurred by the power system and its related expenditures. Thus, the batteries will function in such a way that they will be charged at lower power usage and shall act as a power-generating block during peak load hours. Particle swarm algorithm (PSO) is used since this proposed model is complex and nonlinear.

The primary motive is to find a way that could reduce the electricity bill and also witness reduced peak-average ratio (PAR). Ensuring minimal waiting time for users is also important. Hence, in [27], an efficient home energy management scheme using the meta-heuristic GA, crow search algorithm (CSA), and cuckoo search optimization algorithm (CSOA) is proposed, which can be used to reduce cost and peak load along with a less customer's waiting time. Integrating smart ESS is necessary to ensure efficient operation of HEMS.

The idea of prioritizing by decreasing and scheduling energy consumption is the backbone for all HEMSs. In order to well balance demand and supply, residential demand response program is offered by many utilities so that the graph patterns of energy usage of a user are changed by shifting their energy consumption in peak load hours. In [28], to manage the power consumption, a real-time optimal schedule controller for HEMS using a new binary backtracking search algorithm (BBSA) is suggested. This algorithm provides an optimal schedule for all the appliances used in household in order to reduce total load by scheduling these appliances at different and focused times during daytime.

Reducing the cost using wireless communication is an innovative idea. In [29], an indoor positioning system that is cost-efficient for efficient use of energy in houses is proposed. This method aims at locating position of non-line-of-sight environments using KNN algorithm implemented using Wi-fi through two modes (online and offline) in smartphones. This indoor positioning system locates radio map using noise covariance, and 40 percentage of better results were observed.

The moth flame optimization (MFO) algorithm and GA are two bio-inspired heuristic algorithms, which are used for EMS in smart buildings as proposed in [30]. Their cost reduction rate decreases in PAR value, and comfort level of users is critically evaluated and discussed. Time constrained genetic-moth flame optimization (TG-MFO) algorithm, which is a hybrid version combining GA and MFO, has been developed in order to achieve the energy goals. This algorithm has time constraints per appliance so as to ensure maximum customer comfort. It is carefully designed aiming to eliminate end user discomfort by considering the duration required for each appliance, thus scheduling it properly in response. Renewable energy sources and battery storage units are integrated in order to achieve maximum consumer benefits. Five bio-inspired heuristic algorithms, i.e., firefly algorithm (FA), MFO, GA, ant colony optimization (ACO), and cuckoo search algorithm (CSA), are used to achieve these objectives in living areas by comparing with TG-MFO.

As we witness technological advances in renewable sector and simultaneous reduction in cost, microgrids are slowly entering the energy sector as a means of distributed generation. It can be incorporated for both isolated residential sectors or can also be joined with the already existing electricity grid. Reference [31] presents a methodology, which focuses on two goals, namely decrement in computation time for real-time operations and decrement in everyday functioning cost of microgrid by relaxing method of reducing integer variables.

In [32], the suggested EMS is based on forecasting for energy system, which helps to run a particular defined algorithm. This rolling time horizon (RTH) strategy proves to be extremely effective when the performance of the prediction system is good. Various lightweight models of EMS were designed using machine learning (ML) algorithms and have been compared accordingly considering scenarios of six different simulations.

In [33], for smart home energy management, a hybrid robust-stochastic optimization model is developed. Day-ahead (DA) and real-time (RT) energy markets with fluctuating electricity prices and PV generation are analyzed in this model. When PV generation is worst along with the uncertainties of prices, a flexible robust optimization algorithm (ROA) is created to manage the problem.

Using a two-way communication between each set, a self-adapting Central Energy Management System (CEMS) is developed, which offers a control box capability of adapting and efficiently functioning with any Hybrid-Micro Grids (H-MG) along with integrated generation and storage technologies. This system in [34] operates based on the data parameterization like the amount of non-responsive load demand, the available power from renewable energy resources, and timely scheduling for a range of integrated generation and demand units and the wholesale offers from generation units.

The developed EMS in [35] has day-ahead and hour-ahead scheduling, and concurrent economic dispatch while functioning in real time. Dual stages, management of t power generated from sources in both Micro Grids (MG), and controllable load management are used to achieve optimal commitment. While comparing the traditional analytic methods for power system economic dispatch, meta-heuristic algorithms are used to manage the task of optimal scheduling as they are more suitable. Porcellio Scaber algorithm (PSA) that offers improved efficiency in reduction of objective function is applied on its modified version to solve the optimization problem.

A global optimization technique, which is branch and reduce optimization navigator (BARON), is suggested in [36]. Operation at limited emission and cost is termed as a multi-objective case, while reduction in cost and emission cases is single-objective problem. To optimize this multi-objective case via constraint approach, the results of the single-objective cases are used to calculate possible constraints.

By taking into consideration the flexibility of contracted power of each department, an EMS is developed in order to decrease the power demand of residential buildings. The external grid is supplemented by microgrids – battery energy storage systems (BESS), PV generation panels, and EVs to balance demand and supply. In [37], a mixed binary linear programming (MBLP) formulation is designed in order to optimize the scheduling of the charge and discharge processes of the electric vehicles.

Reference [38] presents a mixed integer nonlinear programming model to optimize the functioning of multiple buildings in a microgrid. This model plans to reduce the total expenditure on energy obtained from the main grid at the interconnection point, balancing generation of buildings, and power demand. This method considers the management of HVAC, lighting appliances, PV generation, and ESS for every individual building. Following a set of linearization techniques and equivalent representation obtained from preprocessing stage worked in EnergyPlus software, a new strategy that simplifies original model is presented. This model was tested in a 13-bus microgrid for various cases of study with non-manageable loads and smart buildings. A large-size test case is also examined with a RTH strategy being planned with the goal of focusing on data uncertainty, as well as reducing the requirement of the amount of forecasting data.

Reference [39] considers a control algorithm that can handle characteristics by using hybridization layers between the weight of neural networks and memory of physical parameters. In order to achieve fast tuned operation, the application of nonlinear regression to the offline hybrid layers construction and online fine-tuning methods is performed using the Gauss–Newton method. The feedforward strategy is used to improve the stability of the overall system. The proposed control performance results are examined and compared to hybrid PID cascade control using the HVAC system. These results showed that feedforward hybrid layer control (FHLC) was more advantageous when it came to adaptation, precision, optimal performance, and robustness.

15.4   IoT AND SMART TECHNOLOGIES

As electricity costs continue to rise and energy resources become scarce, it is becoming more and more important for everyone to be aware of how to utilize energy most efficiently. Smart technologies like Internet of Things (IoT) and big data have found a way for homes and businesses to do this, therefore promoting energy conservation to a large extent.

One proposed method in [40] is to develop an advanced IoT system for integrating effectively with the information ICT sector. This system combines all devices and produces information on the user end, such as the amount of energy consumed, load shifting schedules, calculation of their impact, and conservation using real-time data such as consumption and occupancy measured by sensors, which are placed in the buildings.

In order to optimize building energy management (BEM), it is important to have a flexible model capable of scheduling appliances that are flexible in an optimal manner resulting in less wastage of energy. To make this possible, an IoT system that can analyze the pattern of energy wastage and usage can be used, as proposed in [41].

Future buildings will be relying more on systems that are connected to a network system comprising actuators and sensors. Therefore, in order to enable new potentials pertaining to the development of applications, a fully dedicated system has to be included for the processing and data storage of sensors [42]. The efficiency of a wireless monitoring system for optimal occupancy detection for energy management in smart buildings can be easily integrated with existing BEMS, using sensors. Passive Infrared Sensors (PIR) are used to reduce energy usage, including a sensing hole mechanism that can find places in the varying range of PIR. MILP optimization algorithm is used to monitor and manage energy usage efficiently and to reduce both building energy wastage and system energy wastage simultaneously [43].

To achieve energy efficiency in smart buildings, it is also important to have active engagement from the occupants. Reference [44] proposes an information technology (IT) ecosystem, which focuses on improving energy efficiency in smart buildings by incorporating behavioral change in occupants. ICTs like data modeling and IoT are combined to give an approach to exploit the energy, environmental, and behavioral data, thus leading to EMS according to the user's demands [44].

The smart metering system (SMS) by using real-time monitoring system has modernized the power grid system using real-time monitoring systems resulting in creation of many EMS opportunities. The proposed system models a dynamic EMS for residential buildings connecting a microgrid to a grid employing an SMS for collecting data from various parts of the electrical system. Therefore, in this process of the EMS, the energy storage system plays a vital role. Implementation of strategies pertaining to energy management is allowed by intelligent metering [45].

In [46], by deploying "smart meter," which is a monitoring system, utility metering is integrated into residential management systems. Energy management in large, multi-site locations is often heterogeneous, where responsibility of daily operation and maintenance is dispersed across many individuals. Through this, a large number of users get impacted directly. The users have very little vision about the usage of water and energy. The proposed system for monitoring of the systems aids in accessing the information in a much better way, thus supporting smoother communication, and paves way for better practices [46].

With the aim of increasing efficiency of energy and saving even more energy, this paper [47] proposes the controlling plan and strategy for efficient EMS in smart microgrid and development of an advanced demand-side management. Major challenges faced by Smart Micro Grids (SMG) are CO_2 emissions, voltage and frequency regulations, and load dynamics, which can be overcome by new control and management approaches. The system of demand-side management is used so that the cost of emission, the cost of energy usage, and PAR are lessened. IoT can be used to ensure that the smart grid system is operating in a secured manner and to adaptively monitor the usage of energy, by communication along the generating units, the loads, and the main controller unit. The controller here can help in regulating the parameters like the frequency and voltage of the smart grid system so that the required values are attained at the operating conditions and the defined load demands. With the advent of smart appliances and intelligent systems, it is also of high value to strike a balance between energy consumption and user comfort. Therefore, a Fuzzy Inference System integrated with the IoT operating system is proposed in [48], which uses humidity as one of the input parameters so that thermostat set points are maintained according to the consumer comfort. Energy usage needed for a medium-sized room with a heater can be lowered by including a feedback loop with the temperature of the room as the input parameter to the Fuzzy Inference System (FIS). We can expand the same later through incorporating the given fuzzy controller to many other houses and further integrate it [48].

It is the alarming need to recognize the large impact caused by the usage of energy in buildings. Energy management of the buildings is identified as an efficient system for maintaining the proper and efficient usage of energy in urban places. This paper [49] presents a method for lowering the energy usage in buildings by identifying and monitoring the required parameters, thereby allowing us to propose a prediction model working at the data at hourly basis in buildings. For generating such models, some computing techniques are applied. For verifying the feasibility, we will take a reference building for which we have data. In order to implement the proposed model, remarking of the building is done with various parameters like the usage of energy of the buildings and other parameters. After this, the most efficient method is selected for generating the appropriate model for the building. This is followed by the prediction of the energy usage by giving some parameters as inputs. Then, considering the building's usage profile, for saving energy, several actions and methods are proposed and discharged.

Smart buildings using ICT enable automated control and operations, therefore enhancing comfort and productivity while using lesser energy than a conventional building. They provide more visibility into the building's operations and energy consumption and are hence a more attractive option nowadays. Reference [50] discusses the different energy efficiency programs offered to conventional buildings to motivate building operations team to make the switch, even in existing buildings. The various techniques used in smart management system of the buildings are described as how they are differentiated from the commonly used methodologies and how much energy is saved at the end. The potential of smart buildings is enormous, which has made many program managers come up with innovative approaches to integrate smart technologies in existing and new program offerings [50].

15.5 CONCLUSION

EMS in microgrids has enabled buildings and industries to control the amount of energy consumption, thereby reducing costs, risk, and carbon emissions. In order to reduce energy usage and lower the costs and emissions due to GHG, several users have aggregated showing a large potential for implementation in smart cities and society. Experimental results showing significant change in cost and carbon impact prove the efficiency of the system.

Smart EMS is more than switching on and off appliances or transferring data. Therefore, a clear definition of the overall goals for the SHEMS is necessary. This transition from systems that are centrally mounted to microgrids is highly recommended in this climate, where resources are becoming scarce. While the futuristic goal is to look for cheap, environment-friendly energy resources, the flexibility of the smart systems opens up an opportunity to integrate, additionally, renewable energy sources like photovoltaic.

REFERENCES

[1] Farmani, F., Parvizimosaed, M., Monsef, H., & Rahimi-Kian, A. (2018), "A conceptual model of a smart energy management system for a residential building equipped with CCHP system", *International Journal of Electrical Power & Energy Systems* 95, 523–536.

[2] Di Piazza, M. C., La Tona, G., Luna, M., & Di Piazza, A. (2017), "A two-stage energy management system for smart buildings reducing the impact of demand uncertainty", *Energy and Buildings* 139, 1–9.

[3] Chauhan, R. K., Rajpurohit, B. S., Wang, L., Gonzalez Longatt, F. M., & Singh, S. N. (2017), "Real time energy management system for smart buildings to minimize the electricity bill", *International Journal of Emerging Electric Power Systems* 18(3).

[4] Meador, J. C., Griffiths, J. C., Sethi, G., Burns, D. W., & Floyd, P. D. (2013), "Commissioning system for smart buildings", U.S. Patent Application 13/661,341.

[5] Kyriakarakos, G., & Dounis, A. (2020), "Intelligent management of distributed energy resources for increased resilience and environmental sustainability of hospitals", *Sustainability* 12(18), 7379.

[6] Haidar, N., Attia, M., Senouci, S.-M., Aglzim, E.-H., Kribeche, A., & Asus, Z. B. (2018), "New consumer-dependent energy management system to reduce cost and carbon impact in smart buildings", *Sustainable Cities and Society* 39, 740–750.

[7] Han, Y., Chen, W., Li, Q., Yang, H., Zare, F., & Zheng, Y. (2018), "Two-level energy management strategy for PV-fuel cell-battery-based DC microgrid", *International Journal of Hydrogen Energy* 44.

[8] Ahmad, F., Alam, M. S., & Asaad, M. (2017), "Developments in xEVs charging infrastructure and energy management system for smart microgrids including xEVs", *Sustainable Cities and Society* 35, 552–564.

[9] Rezaei, N., Mazidi, M., Gholami, M., & Mohiti, M. (2020), "A new stochastic gain adaptive energy management system for smart microgrids considering frequency responsive loads", *Energy Reports* 6, 914–932.

[10] Liu, Y., Qiu, B., Fan, X., Zhu, H., & Han, B. (2016), "Review of smart home energy management systems", *Energy Procedia* 104, 504–508.

[11] Thomas, D., Deblecker, O., & Ioakimidis, C. S. (2018), "Optimal operation of an energy management system for a grid-connected smart building considering photovoltaics' uncertainty and stochastic electric vehicles' driving schedule", *Applied Energy* 210, 1188–1206.

[12] Vergara, P. P., López, J. C., da Silva, L. C. P., & Rider, M. J. (2017), "Security-constrained optimal energy management system for three-phase residential microgrids", *Electric Power Systems Research* 146, 371–382.

[13] Ma, L., Liu, N., Wang, L., Zhang, J., Lei, J., Zeng, Z., ... & Cheng, M. (2016), "Multiparty energy management for smart building cluster with PV systems using automatic demand response", *Energy and Buildings* 121, 11–21.

[14] Indragandhi, V., Logesh, R., Subramaniyaswamy, V., Vijayakumar, V., Siarry, P., & Uden, L. (2018), "Multi-objective optimization and energy management in renewable based AC/DC microgrid", *Computers & Electrical Engineering* 70.

[15] Wang, Y., Li, Y., Cao, Y., Tan, Y., He, L., & Han, J. (2018), "Hybrid AC/DC microgrid architecture with comprehensive control strategy for energy management of smart building", *International Journal of Electrical Power & Energy Systems* 101, 151–161.

[16] Gonçalves, I., Gomes, Á., & Antunes, C. H. (2019), "Optimizing the management of smart home energy resources under different power cost scenarios", *Applied Energy* 242, 351–363.

[17] Querini, P. L., Chiotti, O., & Fernádez, E. (2020), "Cooperative energy management system for networked microgrids", *Sustainable Energy, Grids and Networks* 23, 100371.

[18] Jonban, M. S., Romeral, L., Akbarimajd, A., Ali, Z., Ghazimirsaeid, S. S., Marzband, M., & Putrus, G. (2020), "Autonomous energy management system with self-healing capabilities for green buildings (microgrids)", *Journal of Building Engineering* 101604.

[19] Duman, A. C., Erden, H. S., Gönül, Ö., & Güler, Ö. (2020), "A home energy management system with an integrated smart thermostat for demand response in smart grids", *Sustainable Cities and Society* 65.

[20] Moreno, M. V., Zamora, M. A., & Skarmeta, A. F. (2014), "User-centric smart buildings for energy sustainable smart cities", *Transactions on Emerging Telecommunications Technologies* 25, 41–55.

[21] Hurtado, L. A., Nguyen, P. H., & Kling, W. L. (2014)," Agent-based control for building energy management in the smart grid framework", *IEEE PES Innovative Smart Grid Technologies*, Europe, p. 6.

[22] De Paola, A., Ferraro, P., Re, G. L., Morana, M., & Ortolani, M. (2019), "A fog-based hybrid intelligent system for energy saving in smart buildings", *Journal of Ambient Intelligence and Humanized Computing* 11, 2793–2807.

[23] Rafique, S. F., Zhang, J., Hanan, M., Aslam, W., Rehman, A. U., & Khan, Z. W. (2018), "Energy management system design and testing for smart buildings under uncertain generation (wind/photovoltaic) and demand", *Tsinghua Science and Technology* 23(3), 254–265.

[24] Ostadijafari, M., Dubey, A., Liu, Y., Shi, J., Yu, N. (2019), "Smart building energy management using nonlinear economic model predictive control", pp. 1–5.

[25] Ahmad, A., Khan, A., Javaid, N., Hussain, H. M., Abdul, W., Almogren, A., … Azim Niaz, I. (2017), "An optimized home energy management system with integrated renewable energy and storage resources", *Energies* 10(4), 549.

[26] Dadashi-Rad, M. H., Ghasemi-Marzbali, A., & Ahangar, R. A. (2020), "Modeling and planning of smart buildings energy in power system considering demand response", *Energy* 118770.

[27] Aslam, S., Iqbal, Z., Javaid, N., Khan, Z., Aurangzeb, K., & Haider, S. (2017), "Towards efficient energy management of smart buildings exploiting heuristic optimization with real time and critical peak pricing schemes", *Energies* 10(12), 2065.

[28] Ahmed, M. S., Mohamed, A., Khatib, T., Shareef, H., Homod, R. Z., & Ali, J. A. (2017), "Real time optimal schedule controller for home energy management system using new binary backtracking search algorithm", *Energy and Buildings* 138, 215–227.

[29] Borhani Afuosi, M., & Zoghi, M. R. (2020), "Indoor positioning based on improved weighted KNN for energy management in smart buildings", *Energy and Buildings* 212, 109754.

[30] Ullah, I., & Hussain, S. (2019), "Time-constrained nature-inspired optimization algorithms for an efficient energy management system in smart homes and buildings", *Applied Sciences* 9(4), 792.

[31] McIlvenna, A., Herron, A., Hambrick, J., Ollis, B., & Ostrowski, J. (2020), "Reducing the computational burden of a microgrid energy management system", *Computers & Industrial Engineering* 143, 106384.

[32] Leonori, S., Martino, A., Frattale Mascioli, F. M., & Rizzi, A. (2020), "Microgrid energy management systems design by computational intelligence techniques", *Applied Energy* 277, 115524.

[33] Akbari-Dibavar, A., Nojavan, S., Mohammadi-Ivatloo, B., & Zare, K. (2020), "Smart home energy management using hybrid robust-stochastic optimization", *Computers & Industrial Engineering* 143, 106425.

[34] Marzband, M., Ghazimirsaeid, S. S., Uppal, H., & Fernando, T. (2017), "A real-time evaluation of energy management systems for smart hybrid home Microgrids", *Electric Power Systems Research* 143, 624–633.

[35] Keshta, H. E., Malik, O. P., Saied, E. M., Bendary, F. M., & Al, A. A. (2020), "Energy management system for two islanded interconnected micro-grids using advanced evolutionary algorithms", *Electric Power Systems Research* 192, 106958.

[36] Mosa, M. A., & Ali, A. A. (2020), "Energy management system of low voltage dc microgrid using mixed-integer nonlinear programing and a global optimization technique", *Electric Power Systems Research* 192, 106971.

[37] Foroozandeh, Z., Ramos, S., Soares, J., Lezama, F., Vale, Z., Gomes, A., & Joench, L. R. (2020), "A mixed binary linear programming model for optimal energy management of smart buildings", *Energies* 13(7), 1719.

[38] Pinzon, J. A., Vergara, P. P., da Silva, L. C. P., & Rider, M. J. (2018), "Optimal management of energy consumption and comfort for smart buildings operating in a microgrid", *IEEE Transactions on Smart Grid* 1–1.

[39] Homod, R. Z. (2018), "Analysis and optimization of HVAC control systems based on energy and performance considerations for smart buildings", *Renewable Energy* 126, 49–64.

[40] Marinakis, V., & Doukas, H. (2018), "An advanced IoT-based system for intelligent energy management in buildings", *Sensors* 18(2), 610.

[41] Tushar, W., Yuen, C., Li, K., & Wood, K. L. (2016), "Design of cloud-connected IoT system for smart buildings on energy management", *EAI Endorsed Transactions on Industrial Networks and Intelligent Systems* 3(6).

[42] Linder, L., Vionnet, D., Bacher, J.-P., & Hennebert, J. (2017), "Big building data - a big data platform for smart buildings", *Energy Procedia* 122, 589–594.

[43] Lasla, N., Doudou, M., Djenouri, D., Ouadjaout, A., & Zizoua, C. (2019), "Wireless energy efficient occupancy-monitoring system for smart buildings", *Pervasive and Mobile Computing* 59, 101037.

[44] Fotopoulou, E., Zafeiropoulos, A., Terroso-Sáenz, F., Şimşek, U., González-Vidal, A., Tsiolis, G., & Skarmeta, A. (2017), "Providing personalized energy management and awareness services for energy efficiency in smart buildings", *Sensors* 17(9), 2054.

[45] Mbungu, N. T., Bansal, R. C., Naidoo, R. M., Bettayeb, M., Siti, M. W., & Bipath, M. (2020), "A dynamic energy management system using smart metering", *Applied Energy*, 280, 115990.

[46] Stuart, G., & Ozawa-Meida, L. (2020), "Supporting decentralised energy management through smart monitoring systems in public authorities", *Energies* 13(20), 5398.

[47] Sedhom, B. E., El-Saadawi, M. M., El Moursi, M. S., Hassan, M. A., & Eladl, A. A. (2020), "IoT-based optimal demand side management and control scheme for smart microgrid", *International Journal of Electrical Power & Energy Systems* 127.

[48] Ain, Q., Iqbal, S., Khan, S., Malik, A., Ahmad, I., & Javaid, N. (2018), "IoT operating system based fuzzy inference system for home energy management system in smart buildings", *Sensors* 18(9), 2802.

[49] Moreno, M. V., Dufour, L., Skarmeta, A. F., Jara, A. J., Genoud, D., Ladevie, B., & Bezian, J.-J. (2015), "Big data: the key to energy efficiency in smart buildings", *Soft Computing* 20(5), 1749–1762.

[50] King, J., & Perry, C. (2017), "Smart buildings: Using smart technology to save energy in existing buildings", *American Council for an Energy-Efficient Economy* 46.

16 Augmented Lagrangian Model to Analyze the Synergies of Electric Urban Transport Systems and Energy Distribution in Smart Cities

R. Subramani
Amrita Vishwa Vidyapeetham

C. Vijayalakshmi
Central University of Tamil Nadu

CONTENTS

16.1 INTRODUCTION

This research mainly deals with the design of augmented Lagrangian (AL) model to analyze the energy distribution in smart cities for electric urban transport system [1]. Researchers have analyzed the facilities for transport, but the management of optimizing the energy resources is not much analyzed [2]. An electric power system comprises three major subsystems, namely power generation, transmission, and

DOI: 10.1201/9781003240853-16

189

distribution systems. In the transmission system, the electrical power generated by the generators is consumed by the loads at the consumer side. The transmission lines are operated at loadings below their thermal ratings for maximizing the reliability of power circuits. In the present study, the optimization model is designed for power generation with various parameters and generating conditions of the power system [3]. The objective is to maximize the number of units generation with minimum power production costs in such a way that the expected demand is satisfied. The main assumption is that part of regenerative braking energy can be stored in electric vehicles batteries such that it can be utilized later for other transport systems. Decomposition algorithms are analyzed with respect to various parameters and conditions. The classical method Lagrangian relaxation (LR) is implemented in many combinatorial optimization problems; it has been recently implemented in many inference problems.

LR technique decomposes the optimization problem into subproblems; Lagrangian subproblems give the optimal solution for the optimization problem [4]. The main objective is to identify the controlled and uncontrolled parameters of the various decomposition techniques, which are framed as equality and inequality constraints. More specifically, the objective function is transformed into a Lagrangian problem by adding Lagrangian, which is acting as "penalty factor," based on the system parameters [5]. It is compared with the other decomposition techniques such as primal decomposition and dual decomposition. An optimization model for the plan of distributed energy resources considering the transport system of electric vehicles is designed. Various synergies are being analyzed [6]. Mathematical model is formulated using AL technique. Based on the numerical computations and its graphical representations, the power cost is optimized, and hence, Lagrangian technique can be used for optimizing the time and fuel cost.

The general mathematical problem can be stated as follows:

Minimize $f(x)$

Subject to

$$g_i(x) \geq 0 \; (i = 1, 2, 3, \ldots m) \tag{16.1}$$

$$h_i(x) \geq 0 \; (j = 1, 2, 3, \ldots m) \tag{16.2}$$

$$x \in S \tag{16.3}$$

where $x = (x_1 \, x_2 \, x_3 \ldots x_n)^T$ is the vector of unknown decision variables, and $f, h_j \, (j = 1, 2, 3, \ldots m)$, $g_i \, (i = 1, 2, 3, \ldots m)$ are the real-valued functions of the n real variables $x_1, x_2, x_3, \ldots x_n$, where f is defined as the objective function, and inequalities equations (16.1–16.3) are referred to as the constraints. The mathematical programming problem (MP) is designed as a minimization problem. This has been done without any loss of generality by using the identity:

$$\text{Max } f(x) = -\text{Min } (-f(x)) \tag{16.4}$$

Generally, when discussing a problem of the type (MP), assumptions, in addition to those stated, on the functions are made so that the problem is more tractable to theoretical treatment. For example, the functions f, g_i, h_j are usually assumed to be continuous or continuously differentiable. Also, the set S is normally taken as a connected subset of R^n. This enables us to make small changes in x. In general, S is considered as the entire space R^n.

Let T be the set of all points x, which satisfy constraints equations (16.1)–(16.3). Then, the set T is said to be the feasible region, feasible set, or constraint set of problem (MP). A feasible solution is an n-dimensional vector with all the constraints of (MP). If the constraint set T is empty, then there is no feasible solution; in this case, program (MP) is inconsistent [7]. A feasible point x^0 of program (MP), i.e., $x^0 \in T$, is known as an optimal, a global optimal, or an absolute optimal solution to program (MP) if

$$f(x) \geq f\left(x^0\right) \text{for all } x \in T \tag{16.5}$$

A global optimal solution x^0 of (MP) is indeed a global minimum point of problem (MP). Thus, we can say that the function $f(x)$ has a global minimum at $x^0 \in T$ if inequality equation (16.5) holds. A point $x^0 \in T$ is said to be strict global minimum point of f over T if the strict inequality ($>$) in equation (16.5) holds for all $x^0 \in T$, $x \neq x^0$. A point $x^0 \in T$ is referred to as a global maximum (or strict global maximum) point of f over T if x^0 is a global minimum (or strict global minimum) point of $-f$ over T. The point $x^* \in T$ is called the local minimum point of f over T if there exists some $\delta > 0$ such that

$$f(x) \geq f\left(x^*\right) \text{for all } x \in T \cap N_\delta\left(x^*\right) \tag{16.6}$$

where $N_\delta(x^*)$ is the neighborhood of x^* and has a radius δ. If the strict inequality ($>$) in relation (16.6) holds for all $x \neq x^0$ and $x \in T \cap N_\delta\left(x^*\right)$, then x^* is known as a strict local minimum point of f over T. A point $x^* \in T$ is called a local maximum point of f over T if x^* is a local minimum (or strict local minimum) point of $-f$ over T. An extremum point is either a local maximum point or a local minimum point. The mathematical programming problem (MP) can be broadly classified into two categories, namely unconstrained optimization problems and constrained optimization problems. If the constraint set T is the whole space R^n, program (MP) is then called unconstrained optimization problem [7]. In this case, we are interested in finding a point of R^n at which the objective function has an optimum value. On the other hand, if T is a proper subset of R^n, program (MP) then represents a constrained optimization problem.

The applications of mathematical modeling optimization techniques have been discussed for operating power systems and scheduling them. Many researchers have analyzed the operation of the power system with respect to generation, transmission, and distribution [8]. The electrical system is designed to plan and manage the increasing electricity demand by minimizing the losses. The main parameters for the power management system are voltages and currents. The power systems are designed by both controlled and uncontrolled parameters of the system [9]. Loads are defined as uncontrolled parameters, and they can vary with time and affect the behavior of the power system. In this context, it is to be mentioned that optimization models preserve the stability and security. In the present study, an optimization technique leads to the solution of real-time problems.

Energy has played a key role in enhancing the quality of life of human beings in the past and will continue to be a critical parameter in deciding the economic and social well-being of the people in the years to come. However, increased and indiscriminate utilization of non-renewable sources of energy has created many problems in different parts of the world, the most significant among them being environmental degradation [10]. A wide cross section of people around the world are now aware of the great danger that is posed due to global warming – the prime effect of greenhouse gases. A great deal of interest has therefore been evinced on the use of naturally available renewable sources of energy to check the further degradation of the environment. World energy demand fuel is depicted in Figure 16.1 and consumptions in Figure 16.2 in which the demand increases over the 2035 period.

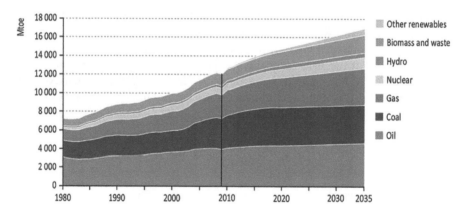

FIGURE 16.1   Word energy demand fuel. (IEA, WEO-2011.)

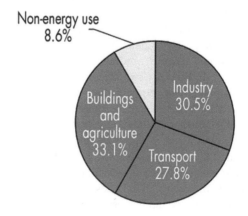

FIGURE 16.2   Consumptions by sector as per new policies 2040. (KWES-2015 and IEA.)

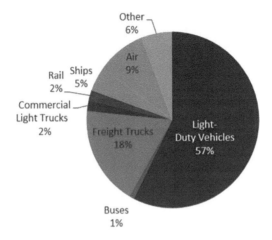

FIGURE 16.3   Transportation energy. (IEA and EIA.)

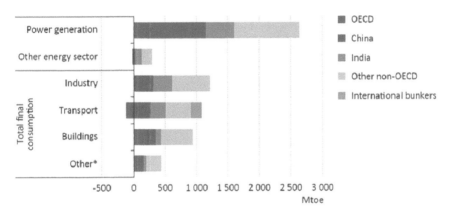

*Other includes agriculture and non-energy use. Note: Total final consumption includes electricity and heat generated by the power sector.

FIGURE 16.4   Energy demand by sector and region in the new policies. (IEA, WEO-2014.)

Transportation energy and change in energy demand by sectors and regions throughout the globe are depicted in Figures 16.3 and 16.4. Sustainability is discussed in Figure 16.5.

16.2   AUGMENTED LAGRANGIAN METHOD

An AL method is implemented to solve this problem [11]. In order to solve this problem, initially, the inequality constraints are transformed to equality constraints by adding slack variables (a). Therefore, the given (primal) problem becomes

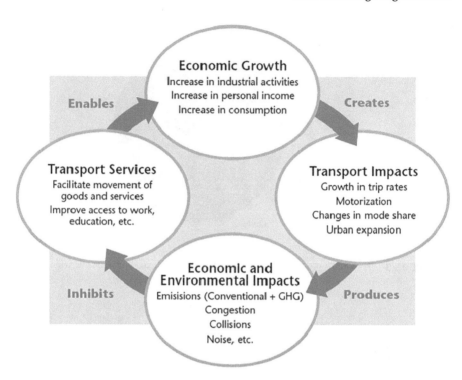

FIGURE 16.5 Challenges of making mobility sustainable. (WBCSD-2004.)

$$\begin{cases} \underset{p,I}{\mathrm{Min}}\, f(p,I) \\[2ex] \qquad \text{Subject to} \\[2ex] \qquad\qquad\qquad g(p,I)\ -\alpha=0 \end{cases} \qquad (16.7)$$

The corresponding AL method is defined by

$$L*(p,I,\lambda,c,\alpha)=f(p,I)+< \lambda,g(p,I)-\alpha >+\frac{c}{2}\|g(p,I)-\alpha\|^2 \qquad (16.8)$$

Here, the inner product of two vectors is defined by $<x, y>=x^T y$,

$\lambda=[\lambda_1\ \lambda_2\ \lambda_3]^T$ as a three-dimensional vector, α is an m-dimensional vector, act as a Lagrangian multipliers associated with system constraints such as demand constraint, emission constraint over a time period in Nt. The slack variables are eliminated by defining

$$L(p,I,\lambda,c)=\underset{\alpha}{\mathrm{Min}}\, L*(p,I,\lambda,c,\alpha) \qquad (16.9)$$

16.3 INITIALIZING THE MULTIPLIERS

The number of iterations of high level can be reduced significantly based on the initialization of multipliers. Multipliers and their system load are initialized based on dispatch and the commitment state. The thermal units are initially committed in a sequential order to fulfill the load (demand), and then, hydro units are allotted for commitment over a planning horizon [1,2]. This process continues until constraint conditions are satisfied. The multipliers λ_i are initialized based on the marginal cost of the system over k hours, and all other multipliers converge to zero.

16.4 PROCEDURE FOR FEASIBILITY

Dual solutions are normally infeasible. To obtain feasible solution, a heuristic method is demonstrated based on the dual solutions. In order to obtain the solutions of the subproblems, dual solutions satisfy energy constraints and capacity. Among all the feasible solutions, the better one is selected as an optimal solution. Based on the solutions of the subproblem, feasibility gap has been calculated, and that leads to the quality of the feasible schedule.

Transmission network [1], its distribution, and sustainable network design are displayed in the below Figures 16.6–16.10.

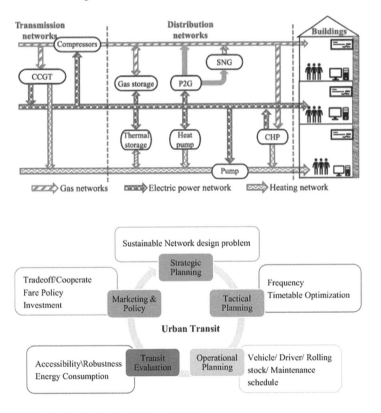

FIGURE 16.6 Centralized scheduling architecture.

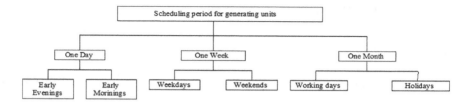

FIGURE 16.7 Scheduling design.

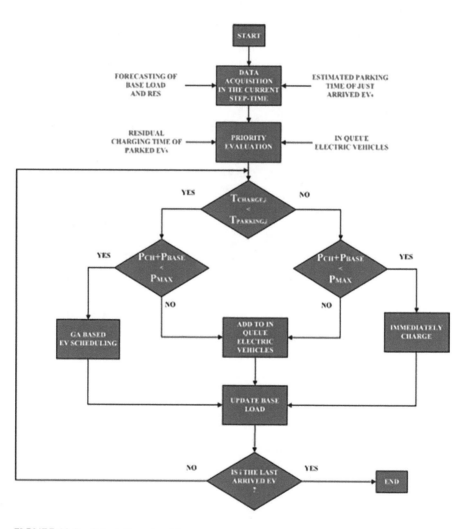

FIGURE 16.8 Scheduling algorithm.

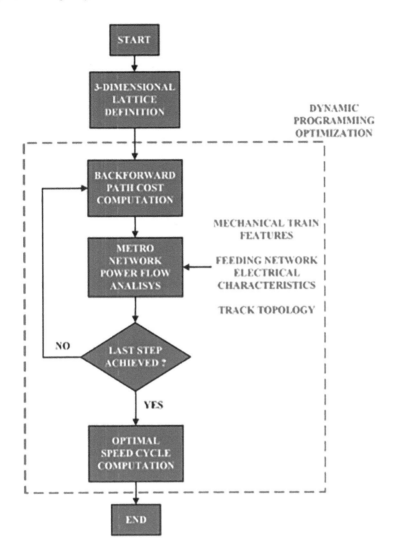

FIGURE 16.9 Dynamic programming algorithm.

16.5 OPTIMIZATION MODEL FORMULATION

Parameters

C_p – cost for power

C_{EE} – electric energy cost for the place

C_{PVi} – photovoltaic investment cost

C_{PVOM} – operation and maintenance cost for photovoltaic

C_{Bi} – battery investment cost

C_{TE} – thermal energy cost for the place

C_{EV} – cost for electric vehicle

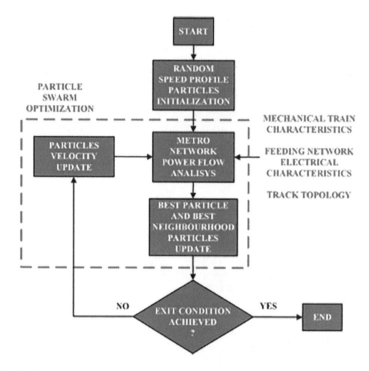

FIGURE 16.10 Particle swarm optimization algorithm.

PV_{DNI} – photovoltaic energy in W/m² (through direct normal irradiance (DNI))
PV – photovoltaic production
PV_p – peak photovoltaic power (installed capacity) in kW
PV_l – losses in photovoltaic in %
G – global irradiation in horizontal plane (theoretically calculated as 1,000 W/m²)
D_E – actual electricity demand in kWh
D_{New} – new electricity demand in kWh
D_{inc} – demand increment per hour in kWh
D_{dec} – demand decrement per hour in kWh
S_l – system losses in %
EV_c – electric vehicle charging point
EV_n – number of electric vehicles
E_c – energy charged in electric vehicle in kWh
E_d – energy discharged from electric vehicle in kWh
EV_s – storage availability of electric vehicle in kWh
λ_i, μ_j – Lagrangian multipliers

16.6 OBJECTIVE FUNCTION AND CONSTRAINTS

The objective is to minimize overall economic cost (total establishment cost (TES)), which is formulated as

$$\text{Minimize } TES = C_p + C_{EE} + C_{PVi} + C_{PVOM} + C_{Bi} + C_{TE} + C_{EV}$$

Subject to the following constraints,

The overall energy efficiency produced by the photovoltaic system can be obtained by

$$PV = \left(\frac{PV_{DNI} * PV_p}{G} \right) * (1 - PV_l) \tag{16.10}$$

Demand response is one of the most important energy features of facilities.

$$\sum D_{New} = \sum D_E \tag{16.11}$$

$$D_{New} = D_E + D_{inc} - D_{dec} \tag{16.12}$$

$$D_{New} \geq 0.1 \text{ kWh} \tag{16.13}$$

The transport system has been modeled with stationary batteries for electric vehicles, which has a set of constraints,

$$EV_c\big|_{h=0} = 0 \tag{16.14}$$

$$EV_c\big|_{m, h=0} = EV_c\big|_{m-1, h=24}, \ \forall m \in \ [2,4] \tag{16.15}$$

$$EV_c\big|_{m, h=0} = EV_c\big|_{m-1, h=24}, \ \forall m \in \ [2,7] \tag{16.16}$$

$$EV_c \leq EV_s * EV_n \tag{16.17}$$

16.7 AUGMENTED LAGRANGIAN MODEL

From the proposed problem, the total cost to be minimized and hence implemented the Lagrangian relaxation technique to exploit the decomposability of the proposed problem. The solution of the subproblems is piecewise linear cost function, which attains the optimal solution only in bounds, and it may oscillate depending on a small change in the multipliers [7]. To overcome this difficulty of Lagrangian, a quadratic penalty function is added with demand constraint and is known as augmented Lagrangian relaxation technique. The AL function for the proposed problem is denoted by

$$L(U, \ \lambda, \ \mu, \ c) = C_{TE} + \sum_i \lambda_i \left[D_E + S_l + PV_l - PV_p \right] +$$

$$\frac{C_i}{2} \sum_i \lambda_i \left[D_E + S_l + PV_l - PV_p \right]^2$$

$$+ \sum_j \mu_j \left[E_c + \alpha_1 - EV_c \right] + \frac{C_j}{2} \sum_j \left[E_c + \alpha_2 - EV_c \right]^2 \tag{16.18}$$

Here, C_i, C_j are positive penalty coefficients and α_1, α_2 are the slack variables. The quadratic penalty terms in equation (16.12) are relaxed by Lagrangian decomposition.

16.8   AUGMENTED LAGRANGIAN RELAXATION ALGORITHM

The algorithm is summarized in the following manner:

1. Initialize the demand multipliers λ_i for the system based on commitment, priority list, and dispatch.
2. Initialize all other multipliers and set the value to zero.
3. Subproblems are solved by using the generated value of multipliers (λ and μ). If the thermal units are present in the algorithm, then, go to Step 3a; otherwise, go to Step 3b.

   3a. Thermal subproblems are solved simultaneously.

   3b. Solve the photovoltaic subproblems simultaneously.
4. Multipliers λ and μ are updated based on the proposed technique.
5. Continue steps 1–4 until stopping criteria are attained; otherwise, go to Step 4.
6. If the obtained solution is feasible, with respect to the commitment state, go to Step 6a; otherwise, go to Step 6b.

   6a. Generated solution by economic dispatch. Go to Step 7.

   6b. Obtain more than one feasible solution using the discussed method.
7. Choose the appropriate feasible solutions, if it reaches the desired number of iterations.
8. Compute higher-order heuristic algorithms based on steps 1–6.

16.9   NUMERICAL CALCULATIONS

By using real-time data, LR gives the optimized results as per the table given below. A comparative analysis is made among PSO, DPA, and augmented Lagrangian relaxation algorithm (ALRA) in which Lagrangian gives the optimized fuel cost and the computational time is 14.23 seconds, which leads to an optimal solution (Table 16.1).

TABLE 16.1

Comparative Analysis of Optimization Algorithms

Method	Total Fuel Cost (in $)	Computational Time (in seconds)
Particle swarm optimization (PSO)	27,878.35	22.32
Dynamic programming algorithm (DPA)	27,865.88	21.14
Augmented Lagrangian relaxation algorithm(ALRA) – proposed method	24,032.70	14.23

16.10 CONCLUSION

AL model is designed to analyze the energy systems within cities, and potential strategies are derived. Algorithm is designed, which helps to recover the energy. The synergies between systems are analyzed through different constraints. These constraints are implemented with respect to different sources in terms of thermal energy, photovoltaic energy, and battery energy for electric vehicles [12–15]. LR technique is a powerful tool, which helps for optimizing the cost. The proposed model can be customized to include different situations and to be projected in number of urban contexts, which can also be adapted to contribute in different energy tariffs. The numerical analysis allows to get more potential benefits of implementation of smart cities in the present scenario.

The interactions between urban energy systems have to be analyzed. This model leads to the reduction of global cost. The main objective is to analyze the different costs such as electric energy, power cost for that location, photovoltaic investment, operation and maintenance cost, battery, thermal energy and battery degradation have to be optimized. The overall energy efficiency is maximized. An algorithm is designed, and feasible solution for economic dispatch is arrived, which helps the society in this current scenario.

REFERENCES

[1] Abdmouleh Z, Alammari RAM, Gastli A, "Review of policies encouraging renewable energy integration & best practices", *Renewable Sustainable Energy Rev.* (2015), 45, 249–262.

[2] Tan Z, Chen K, Ju L, "Issues and solutions of China's generation resource utilization based on sustainable development", *J. Mod. Power Syst. Clean Energy* (2016), 4, 2, 147–160.

[3] Li Z, Al Hassan R, Shahidehpour M, "A hierarchical framework for intelligent traffic management in smart cities", *IEEE Trans. Smart Grid* (2019), 10, 1, 691–701.

[4] Grunditz E, Thiringer T, "Performance analysis of current BEVs based on a comprehensive review of specifications", *IEEE Trans. Trans. Electric.* (2016), 2, 3, 270–289.

[5] Kalathil D, Wu C, Poolla K, "The sharing economy for the electricity storage", *IEEE Trans. Smart Grid* (2019) 10, 1, 556–567.

[5] Yong J, Ramachandaramurthy V, Tan K, "A review on the state-of-theart technologies of electric vehicle, its impacts and prospects", *Renewable Sustainable Energy Rev.* (2015), 49, 365–385.

[6] Xiang Y, Liu Z, Liu J, "Integrated traffic-power simulation framework for electric vehicle charging stations based on cellular automaton", *J. Mod. Power Syst. Clean Energy* (2018), 6, 4, 816–820.

[7] Liu P, Yu J, "Identification of charging behavior characteristic for large-scale heterogeneous electric vehicle fleet", *J. Mod. Power Syst. Clean Energy* (2018), 6, 3, 567–581.

[8] Kwasinski A, Bae S, "Spatial and temporal model of electric vehicle charging demand", *IEEE Trans. Smart Grid* (2012), 3, 1, 394–403.

[9] Shenzhen News Website Charging stations out of service caused large-scale outage of taxis in Shenzhen, http://news.sznews.com/content/2018-05/22/content_19164366. html, Accessed 15 November 2018.

[10] Galus M, Waraich R, Noembrini F, "Integrating power systems, transport systems and vehicle technology for electric mobility impact assessment and efficient control", *IEEE Trans. Smart Grid* (2012), 3, 2, 934–949.

[11] Xu M, Meng Q, Liu K, "Network user equilibrium problems for the mixed battery electric vehicles and gasoline vehicles subject to battery swapping stations and road grade constraints", *Trans. Res. B Method* (2017), 9, 138–166.

[12] Wang SJ, Shahidehpour SM, Kirschen DS, Mokhtari S, Irisarri GD, "Short-term generation scheduling with transmission constraints using augmented lagrangian relaxation", *IEEE Trans. Power Syst.* (1995), 10, 3, 1294–1301.

[13] Rodrigues RN, Finardi EC, da Silva EL, Solving the short-term scheduling problem of hydrothermal systems via lagrangian relaxation and augmented lagrangian, *Math. Problems Eng.* (2012), 2012, 18. doi: 10.1155/2012/856178.

[14] Subramani R, Vijayalakshmi C, "Design of lagrangian decomposition model for energy management using SCADA system", *Smart Innov. Syst. Technol.* (2016), 49, 353–361. doi: 10.1007/978-3-319-30348-2-30.

[15] Subramani R, Vijayalakshmi C, "Augmented lagrangian algorithm for hydrothermal scheduling", *EAI Endorsed Trans. Energy Web Inf. Technol.* (2018), 5, 18, 1–7. doi: 10.4108/eai.12-6-2018.154815.

17 Grid-Interconnected Photovoltaic Power System with LCL Filter Feasible for Rooftop Terracing

V. Meenakshi
Sathyabama Institute of Science and Technology

V. Vijeya Kaveri
Sri Krishna College of Engineering and Technology

G.D. Anbarasi Jebaselvi
Sathyabama Institute of Science and Technology

CONTENTS

DOI: 10.1201/9781003240853-17

17.1   INTRODUCTION

Power interruptions and power blackouts affect the progress of the growing industries by posing more shutdowns and disturb the academic stream in institutes predominantly. In addition, transition from fossil fuels to non-conventional resources aiming for a pollution-free environment leads forcibly to opt for a solar source that is deliberately available in tropical countries such as India.

According to the statistical data available in IISc, Bangalore, it is found that the average solar irradiation in Tamil Nadu is 1,266.52 W/m². Normally, a 1-kWp solar rooftop system can generate an average of 5kWh of electricity/day considering a minimum of five sunshine hours. This installation will be equivalent of planting 20 teak trees over the lifetime and thus reducing carbon dioxide emissions by 15.5 tonnes per year drastically. Fast depletion of fossil fuels and severe radiation from nuclear power plants creating health hazards all have led the researchers to promote renewable energy sources for power production than ever before in these days. Over the last few decades, many renewable energy resources such as biomass, solar, tidal waves, and wind have been found as the best alternatives considering their conversion efficiency and unit cost production. Among these renewable energy sources, solar energy stands as a right substitute to conventional energy sources for electricity production.

The authors discuss different control techniques in particle swarm optimization (PSO), perturb and observe (P&O), and incremental conductance (IC). Modeling is done and simulation is carried out in MATLAB/Simulink [1]. LCL filter is modeled and designed for the front-end converter to reduce current ripple injected to the grid [2]. MPC is a rapid operating controller used to tie the solar system to the grid for actual and quadrature power decoupling [3]. The IC method is used to track maximum power in single-phase grid-connected photovoltaic (PV) systems [4]. The authors used the fuzzy logic control technique and sinusoidal pulse width modulation (SPWM) switching technique for stand-alone PV systems [5]. The authors discussed a10-kW solar panel connected to the grid and also discussed real power injection [6]. The authors described storage devices such as battery for solar rooftop and also estimated the storage capacity of the battery [7]. The harmonics reduction in the load side, solar PV module, and grid side are discussed [8].The authors discussed a new topology that satisfies the IEEE standard so that the current injected to the grid ripples is reduced, thus improving the power quality [9]. Rooftop terracing is discussed by the authors, and control techniques for PV module integration are carried out [10]. The maximum power point tracking (MPPT) controller is a highly efficient DC–DC technology. AC–DC–AC is connected between the solar panel and load to get maximum power. The different maximum power point (MPP) techniques are FLC, P&O, and IC, but the most popular are the P&O and IC methods.

17.2   BLOCK DIAGRAM OF THE PROPOSED SYSTEM

Solar power from the sun is extracted by the solar panel that gives input to the MPPT and boost converter which is a DC input. Figure17.1 shows the block diagram of the proposed system. The major block diagram comprises the proposed solar module,

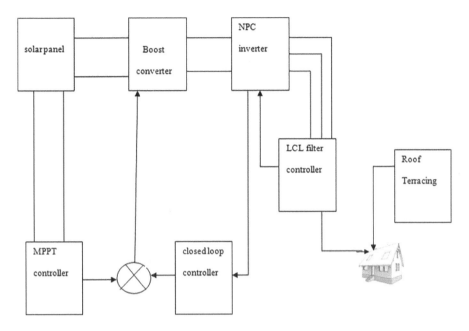

FIGURE 17.1 Block diagram of the proposed system.

boost converter, neutral point clamped (NPC) inverter, PI controller, MPPT controller, LCL filter, and rooftop terracing from the solar panel that is connected with a boost converter and an NPC inverter. The output of this inverter is connected to the closed-loop PI controller. The comparator detects both the outputs coming from closed-loop PI controller and the MPPT control. The output from the LCL filter is connecting to roof terracing. Each block can be explained further in detail as follows:

17.2.1 MPPT CONTROL

The IC algorithm is chosen as the best MPPT algorithm in the simulation of PV systems owing to its rapidity in tracking peak power according to the increasing or decreasing rate of irradiance than all the other methods suggested in the literature. This algorithm compares the panel voltage with the MPPT voltage to get the MPP. It overcomes the disadvantage of the P&O method in tracking the power under the fast-varying atmospheric condition. This method checks for an optimum operating point and reaches the particular MPP and stops disturbing it later. Therefore, it is more proficient than the conventional P&O algorithm. In order to get increased efficiency with limited oscillations while fixing the MPP when it is about to reach, this IC method is greatly recommended.

17.2.2 BOOST CONVERTER

The output from the solar panel is given to a boost converter, and the boost converter increases the output according to laws of conservation energy. It is also called step-up converter. The output from the boost converter is given to the NPC inverter.

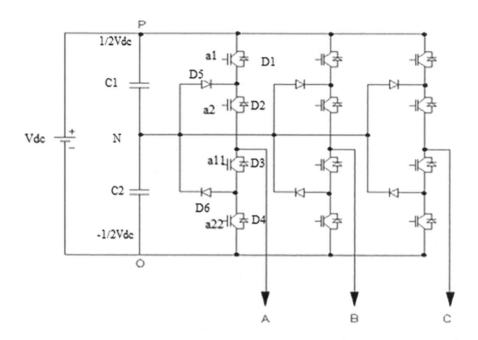

FIGURE 17.2 NPC inverter.

17.2.3 NPC INVERTER

The diode-clamped inverter was also known as the NPC inverter is used in three-stage inverter, (this sentence can be removed). To design an efficient NPC inverter, different ratings of current switches are used. The power switches use different duty cycles. Figure 17.2 shows an NPC inverter. The NPC topology used here has lower ripple in the output current and reduces transients in the output voltage half a time than that of the input, and hence, there is no time delay in the inverter.

17.2.4 CLOSED-LOOP PI CONTROLLER

The input signal proportional and integral $U_i(t)$ is proportional to the output signal $U_o(t)$ and is given by

$$U_o(t) = K_p U_i(t) + K_i \int U_i(t)dt$$

(17.1)

By properly choosing the values of K_p and K_i, the desired output can be obtained. In the PI controller, two error signals are given: one from MPPT and the other signal from PV voltage, which are compared. The PI controller begins to function when the boost converter brings in power from the PV panel and varies the value of the duty cycle.

17.2.5 LCL Filter Controller

In order to reduce current ripple injected to the grid, an LCL filter is used. LCL filters have many advantages; for example, the inductor used in LCL filters is small compared to L filters and are cost-effective. An LCL filter is connected between the inverter and rooftop terracing. The AC output from the NPC inverter is given to the LCL filter to remove the harmonics and produce the desired output from the grid of the system.

17.3 MATERIALS AND METHODOLOGY USED

17.3.1 Descriptive Details of PV Systems

PV cells produce electric power when sunlight or artificial light falls on them. They convert the sun's light energy (photons) into electricity, which either can be transmitted directly or converted to other forms of energy. In the PV module, a thin-film cell coated on silicon is the mostly used material. The cell size is about 10 cm× 10 cm×0.3 mm. In the dark, the PV cell behaves like Shockley diode and the output voltage produced controls the input current of PV cell, and hence, it is called voltage-controlled current source. The interconnected assembly of packaged solar cells constitutes a single PV module or PV panel. A particular PV module can produce certain amount of wattage, and hence, in order to produce large electrical power, several modules or panels need to be connected either in series or in parallel. A complete PV system comprises a group of PV modules, an inverter, batteries, and wiring for interconnection. The ability of these modules at all stages of the PV system development is needed with respect to system sizing, cost analysis, and monitoring.

17.3.2 Materials Used in the PV Cells' Fabrication

The materials used in the fabrication of PV cells are as follows:

Figure 17.3 PV cell shows the most popular desire for solar cells is silicon (Si), with a band hole of 1.1 eV, manufacturing cellular efficiencies of approximately 12%, and a maximum efficiency of about 15%,and gallium arsenide (GaAs), with a band hole of 1.4 eV and a maximum efficiency of about 22%. The maximum theoretical performance for a single mobile is 33%. For multiple cells, the theoretical maximum is 68%. Both of these materials must be grown as single crystals underneath very exactly controlled conditions to limit imperfections, that may purpose recombination. The huge crystals are then sawn to make thin slabs of cell. GaAs is likewise very famous for cells. Gallium and arsenic are precisely one atomic better than silicon, so the machine has many similarities to a silicon-based semiconductor. It is much less friable than silicon, is more resistant to radiation damage, and so is the fabric of choice in space-based cells. Doping this material with atoms from nearby columns within the periodic table adjustments the material as goods piece. In order to absorb light, GaAs is used. Silicon, GaAs, and aluminum gallium arsenide have one-of-a-kind band gaps. They therefore absorb light of different energies. About half the electricity in daylight is unusable via maximum PV cells because this strength is beneath the band gap and so can't free an electron from the valence to the conduction

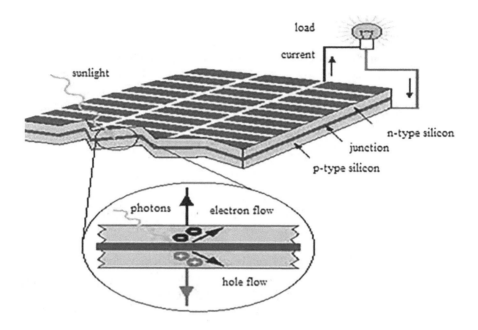

FIGURE 17.3 Photovoltaic cell.

band, or as it carries extra electricity, which ought to be transferred to the cellular as thermal energy, heating up the cell.

17.3.3 PV Cell Characteristics

An equivalent circuit of a solar PV cell consists of a current source illuminated by sunlight and an ideal PN diode (Figure 17.4). When sun light incident on the surface of a PV cell, some portion of the solar energy is absorbed by the semiconductor material. If the absorbed photon energy is greater than the band gap energy of the semiconductor, the electrons bounce from the valence band to the conduction band creating electron–hole pairs throughout the illuminated region of the semiconductor. These electron–hole pairs flow in opposite directions across the junction, thereby creating DC current flow. R_S and R_P are intrinsic series resistance and equivalent shunt resistance. PV cell performance decreases with an increase in temperature, fundamentally owing to increased internal carrier recombination rates, caused by increased carrier concentrations.

Applying Kirchhoff's current law, to the model of PV cell it has been arrived to have the equation as given below:

$$I = n_{pr}I_{ph} - n_{pr}I_{rsc}\left[\exp\left(\frac{q}{KTA} * \frac{V}{n_s}\right) - 1\right] \qquad (17.2)$$

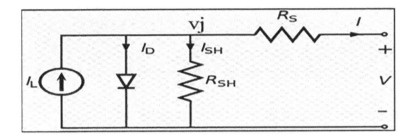

FIGURE 17.4   Equivalent circuit of a solar PV cell.

where
  I– O/P current of PV array
  V–O/P voltage of PV array
  n_{se}– cell series
  n_{pr}– parallel series
  q– charge of an electron
  K– Boltzmann's constant
  A–p–n junction idealistic factor
  T–cell temperature (K)
  I_{rsc}–cell reverse saturation current

Figure 17.5 shows the voltage and current characteristics of a PV cell, and Figure 17.6 shows the short circuit current, MPP, and open circuit voltage of an ideal PV cell.

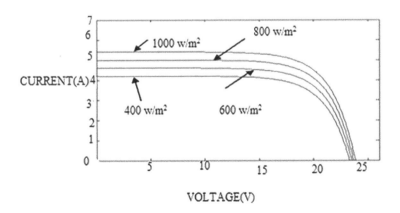

FIGURE 17.5   V-I characteristics of a PV cell.

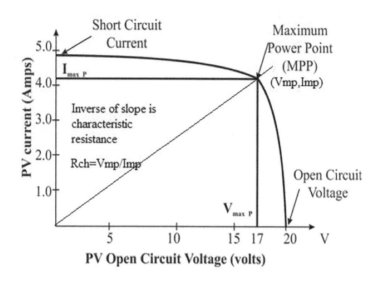

FIGURE 17.6   V-I characteristics of an ideal PV cell.

17.4   WORK PLAN OF ROOFTOP TERRACING

The work plan of this rooftop terracing can be summarized shortly as follows:

- Selection of an existing rural building
- Approval of floor plan of the model unit and setting of a $12' \times 10' \text{ft}^2$ model unit
- Configuring and embedding the PV system between glass panes
- Positioning of the rooftop panels at an angle to capture as much of the incident sunlight as possible
- Employing the following measurements and sensors:
  - An LI-COR pyranometer to measure plane-of-array global solar irradiance(W/m^2)
  - An anemometer to measure wind speed(m/s)
  - A thermistor to measure ambient temperature (°C)
  - An energy meter in the power conditioning unit (PCU)
  - A central computer to receive data from two data acquisition systems
  - Availability of both real-time and archived data on the display along with educational screens.

17.5   SIMULATION RESULTS AND HARDWARE DESCRIPTION

A grid-linked PV electricity system is an energy-producing solar PV energy device linked to the utility grid. Simulations consist of solar panels, a PCU, one or numerous inverters, and the grid connection system. They are used in small residential, business rooftop systems, solar power stations. Unlike stand-alone strength systems, grid-related systems rarely consist of an incorporated battery, as they may be still very luxurious. When situations are proper, the PV system supplies power to load and utility grid. Figure 17.7 shows the Simulink diagram representing the PV module. Figure 17.8 shows the Simulink diagram of the LCL filter. Figure 17.9 shows the Simulink diagram representing the grid-connected system. Figure 17.10 shows the overall simulation diagram of the PV module output voltage and current, Figure 17.11 that of the output voltage at the boost converter, Figure 17.12 that of the load voltage and load current, and Figure 17.13 that of the grid voltage and grid current of the

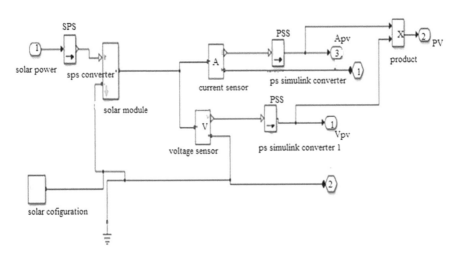

FIGURE 17.7 Simulink diagram representing the PV module.

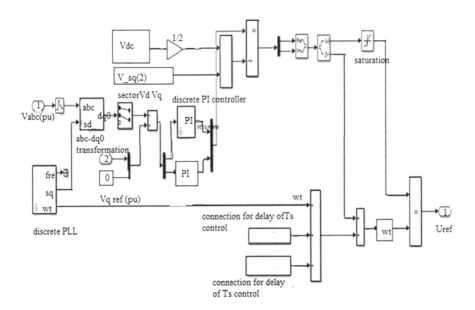

FIGURE 17.8 Simulink diagram of LCL filter.

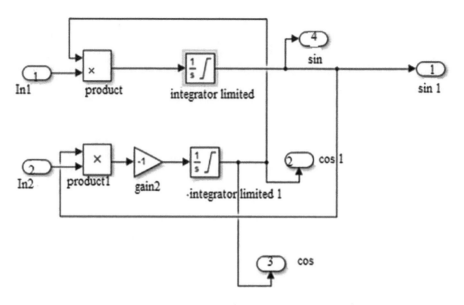

FIGURE 17.9 Simulink diagram representing the grid-connected system.

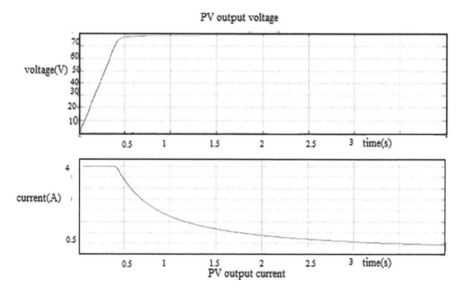

FIGURE 17.10 PV module output voltage and current.

proposed system using MATLAB/Simulink. The simulated output results are shown in the scopes showing PV module output voltage and current, output voltage at the boost converter, load voltage and load current, grid voltage, and load current. It is inferred that the transformer-less inverters suffer higher THD, a remarkable amount of ripples present in grid current, and degradation in the power quality in total, which in turn produces sag and swell in grid current which is indicated at $t = 0.3$ seconds.

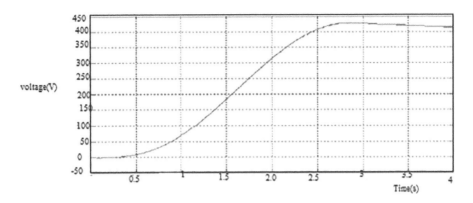

FIGURE 17.11 Output voltage at the boost converter.

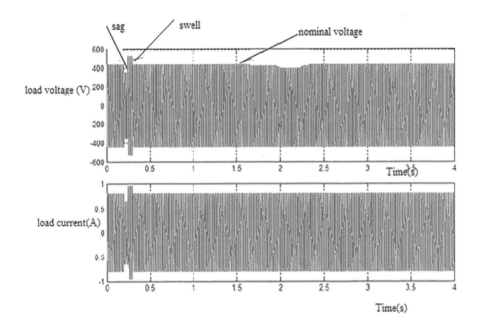

FIGURE 17.12 Load voltage and load current.

Figure 17.14 shows the prototype model of a grid-connected PV system. The voltages at each output have been taken practically by connecting and not connecting the solar panel, respectively. Figure 17.15 model of grid-connected PV system for rooftop terracing shows the whole view of mock-up, horizontal plane. In Table 17.1, the output voltages at various stages are shown. Table 17.2 shows the simulated results for the PV system with MPPT for both the load current and grid current.

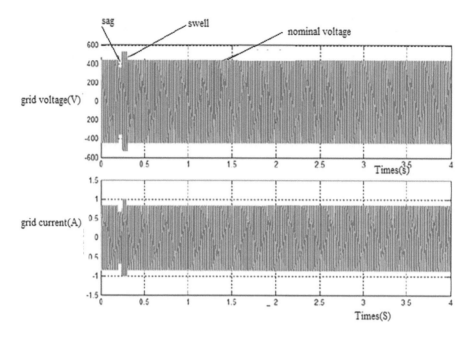

FIGURE 17.13   Grid voltage and grid current.

FIGURE 17.14   Prototype model of the grid-connected PV system.

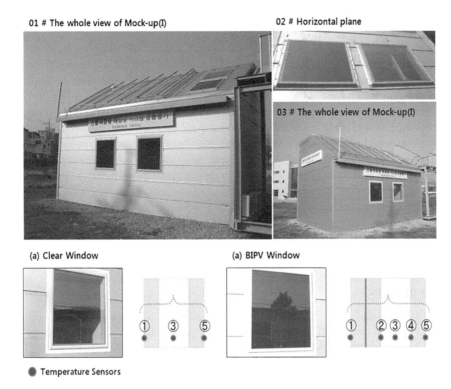

FIGURE 17.15   Model of the grid-connected PV system for rooftop terracing.

TABLE 17.1
Output Voltages at Various Stages

DC Voltage at the Input Side (V)	DC Voltage at the Boost Converter Side (V)	AC Voltage at the Inverter Side (V)
9.5	29.4	22.7

TABLE 17.2
Simulated Results for the PV System with MPPT

Steady State Parameter	Simulated Result
Open circuit voltage	78V
Short circuit current	4 A
Load voltage	425V
Load current	0.75 A
Grid voltage	425V
Grid current	0.75 A
Boost converter voltage	435V

17.6   SUMMARY AND CONCLUSIONS

A grid-connected PV system with MPPT controller and LCL filter has been developed for rooftop terracing. Also, a mathematical model of the PV has been presented. Then, the PV system with a DC–DC boost converter, MPPT controller, LCL filter, and resistive load has been designed.

The PV simulation setup under the MATLAB/Simulink environment has been established by selecting specifications of the PV module and simulated. The IC algorithm has been implemented in MPPT for retrieving optimum solar power to be used in the DC–DC boost converter. Constant voltage is maintained at the output side of the converter. The mitigation of harmonics in grid-connected inverter systems has been affectively done using LCL filter compared to L and LC filters, whenever they are individually used. From the modeling of the boost converter, it was also observed that the output voltage of the boost converter increases along with the increase in duty cycle. A design of stand-alone PV system is analyzed and characteristics of PV system in both online and offline systems are discussed.

REFERENCES

[1] E. A. Gouda, M. F. Kotb, D. A. Elalfy, "Modelling and performance analysis for a PV system based MPPT using advanced techniques", *EJECE, European Journal of Electrical and Computer Engineering* 3, 1,2019.

[2] V. Meenakshi, S. Paramasivam, "Design of LCL filter in front end converter suitable for grid connected wind electric generators", *Journal of Scientific and Industrial Research* 78(12), 896–899, 2019.

[3] A. I. M. Ali, E. E. M. Mohamed, A.-R. Youssef, "MPPT algorithm for grid-connected photovoltaic generation systems via model predictive controller", *2017 Nineteenth International Middle East Power Systems Conference (MEPCON)*, 2017.

[4] G. Shilpa Patil, H. D. Murkute, R. M. Bhombe, "Single phase grid connected PV system using MPPT controller", *IJCMES* 1, 2454–1311, 2017.

[5] V. Subramanian, S. Murugesan, "An artificial intelligent controller for a Three Phase Inverter based Solar PV system using boost converter", *ICOAC* 1, 1–7, 2012, doi: 10.1109/ICoAC.2012.6416840.

[6] S.S. Dheepan, V. Kamaraj, "Grid integration of 10 kW solar panel", *2016 3rd international conference on Electrical Energy Systems (ICEES)*, 2016.

[7] M. Upasani, S. Patil, "Grid connected solar photovoltaic system with battery storage for energy management", *2018 2nd International Conference on Inventive Systems and Control (ICISC)*, 2018.

[8] N. Phannil, C. Jettanesan, A. Ngaopitakkul, "Power quality analysis of grid connected solar power inverter", *2017 IEEE 3rd International Future Energy Electronics Conference and ECCE Asia(IFEE C2017-ECCE Asia)*, 2017.

[9] S. Jain, V. Agarwal, "A single-stage grid connected inverter topology for solar PV systems with maximum power point tracking", *IEEE Trans. Power Electron.* 22, 5, 1928–1940, 2007.

[10] M. Kumar, "Technical issues and performance analysis for grid connected PV system and present solar power scenario", *2020 International Conference on Electrical and Electronics Engineering (ICE3)*, pp. 639–645, 2020.

18 An Efficient ZCS-Based Boost Converter for Commercial Building Lighting Applications

Kishore Eswaran
University of Colorado Boulder

Vinodharani M
KPIT

Srimathi R
Vellore Institute of Technology

Gomathi V
SASTRA

Jaanaa Rubavathay S
Saveetha School of Engineering

Srirevathi B
Vellore Institute of Technology

CONTENTS

DOI: 10.1201/9781003240853-18

18.1 INTRODUCTION

Lighting design [1] and its application in buildings use the illumination of lights to attain an aesthetic effect. Proper lighting design [2] is required to enhance the energy savings, safety, security, atmosphere, visibility, mood, and task performance for the occupants in buildings. Lighting can be classified as indoor lighting and outdoor lighting. Different lamps [3] such as fluorescent, compact fluorescent, neon, halogen, and LED are available for lighting in buildings. Among these, LED lamps have high efficacy, long lifetime, and reliability.

The per capita energy consumption in India is 1,208 kWh [4], of which the commercial building sector consumes 30% of the total energy consumption, i.e., 362.4 kWh. Lighting and air-conditioning in commercial buildings consume nearly 80% (289.92 kWh) of the electrical energy supplied to the commercial sector. Most of the loads in buildings are DC loads. These loads include cell phones, laptops, data centers, electric vehicle charging stations, and lightings. Hence, to power these loads, multiple power conversion stages from the existing AC to DC are employed. To conserve energy, low-voltage DC (LVDC) distribution [5,6] is a prominent attribute, and it is considered as the power source for the lighting loads in this chapter. DC–DC LED drivers are classified as isolated and non-isolated DC–DC converters. Isolated DC–DC converters [7] are bulky and their efficiency is lesser, whereas non-isolated DC–DC converters are efficient and occupy lesser space. Moreover, non-isolated DC–DC converters are sufficient [8,9] for lighting in commercial buildings. The conventional boost converter (BC) has substantial input current ripple and operates at a higher duty ratio with higher switching loss. To enhance the performance of the boost converter, a passive snubber-based boost converter (SBC) is proposed [10,11]. The converter in Ref. [11] is a SBC, and it is considered as LED driver for lighting design in this chapter. The performance of SBC is compared with that of the conventional BC in terms of efficiency and energy savings. A new lighting design is proposed for a tennis court in a commercial building using LED lamps. The proposed design is suggested for the retrofit of High-Pressure Sodium (HPS) lamps. The output lumens of LED lamps in DIALux design is equivalent to the lumens of SBC. Later, the lumen required for tennis court, the lights required, the payback period, and the energy saved by LED lamps compared to HPS lamps will also be presented.

This chapter is organized as follows: A description of circuit description of the SBC including the steady state analysis and the design is given in Section 18.2. Experimental results are examined in Section 18.3, whereas the lighting design using DIALux software for a commercial building space is discussed in Section 18.4. The energy estimation and savings are projected in Section 18.5. Finally, conclusions are presented in Section 18.6.

18.2 CIRCUIT DESCRIPTION

The circuit diagram of the SBC [11] considered for lighting design is given in Figure 18.1.

The converter consists of a power switch S, one main inductor L, two snubber inductors L_{sa} and L_{sb}, output capacitor C, snubber capacitance C_{sn}, main diode D,

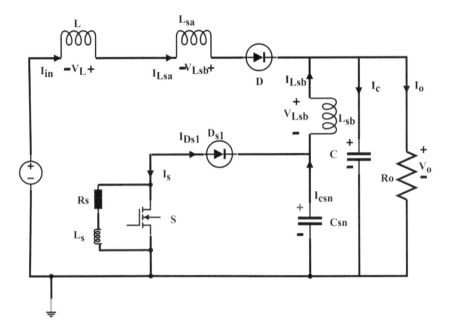

FIGURE 18.1 Snubeer Based Boost Converter.

and a snubber diode D_{s1}. The switch S is operated with a switching frequency of f_s. The duty ratio of the switch is D_T. The following assumptions are considered for the steady state operation of the converter:

1. All the components in the circuit are ideal and the on-state resistance of the switches, forward voltage drop across the diode, and other equivalent series resistances are ignored.
2. The output capacitor C is adequate to maintain a constant voltage across it.
3. The main inductor, L and the input voltage, V_{in} is considered as constant current source, whereas the output voltage V_o is approximated as a constant voltage.

18.2.1 STEADY STATE ANALYSIS

The SBC is operated in a continuous conduction mode (CCM) with five modes of operation for a period of T_s. The switch S is operated with a duty ratio of D_T. The steady state waveforms of the SBC are shown in Figure 18.2. The equivalent circuits for the five modes of operation are shown in Figure 18.3:

1. Mode I: $[t_o–t_1]$: During this mode, the switch S is switched "ON" with zero-current switching (ZCS). The current through the diode D_{s1} (I_{Ds1}) decreases linearly and reaches zero because the voltage across L_{sa} is a constant voltage V_o. The input current I_{in} is constant and is equal to the sum of I_s and I_{Ds1} at t_1.

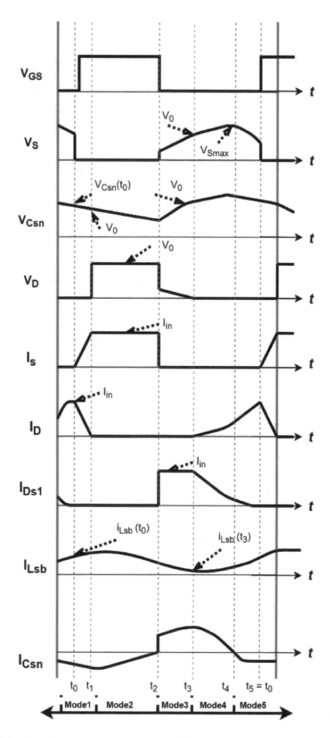

FIGURE 18.2 Steady state waveforms of the SBC.

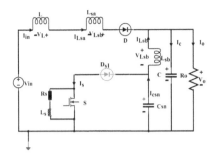

(a) Mode 1: switch S, diode
D - ON, and Ds1 - OFF

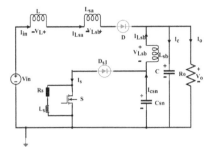

(b) Mode 2: D_{s1}, D – OFF,
and S – ON

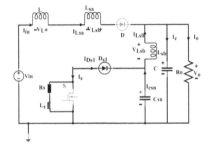

(c) Mode 3: switch S, diode D – OFF
and D_{s1} -ON

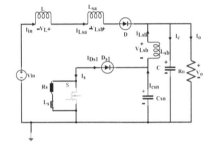

(d) Mode 4: switch S – OFF, Diodes
D_{s1} and D are OFF

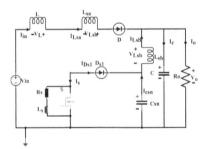

(e) Mode 5: switch S, diodes D_{s1} and D are OFF

FIGURE 18.3 Modes of operation of the SBC.

The capacitor C_{sn} discharges through L_{sa} and reaches a voltage V_o. The current through the switch Q and the diode D during this mode is given in equations (18.1) and (18.2), respectively.

$$i_s(t) = \frac{V_o\,(t - t_o)}{L_{sa}}$$

(18.1)

$$i_D(t) = I_{in} - \frac{V_o\,(t - t_o)}{L_{sa}} \tag{18.2}$$

where I_S and I_D the current through the main switch S and the diode D, V_o is the output voltage, and L_{sa} is the snubber inductance.

Mode II: $[t_1 - t_2]$: In mode II, the snubber diode D_{s1} remains OFF and the main diode D is turned OFF. The main switch S is still ON as in mode I. The voltage across C_{sn} and the current through the L_{sb} at the end of this mode are $V_{Csn}(t_2)$ and $I_{Lsb}(t_2)$, respectively.

Mode III: $[t_2 - t_3]$: During this mode, the switch S is turned OFF and the diode D remains in the OFF state. The diode D_{s1} is turned ON, and therefore, the diode current $I_{Ds1} = I_{in}$ is equal to the sum of I_{Csn} and I_{Lsb}. The input current I_{Csn} is given in equation (18.3):

$$i_{Csn}(t) = \frac{V_o}{Z_{1 - V_{csn}(t_2)}}\sin\omega_1(t - t_2) - i_{Lsb}(t_2 - I_{in})\cos\omega_1(t - t_2) \tag{18.3}$$

$$I_{Lsb} = I_{in} - I_{Csn} \tag{18.4}$$

where $Z_1 = \sqrt{\dfrac{L_{s2}}{C_s}}$ and $\omega_1 = \dfrac{1}{\sqrt{L_{s2}C_s}}$.

The voltage V_{Csn} is increased by the charging current I_{Csn}. The voltage across the switch is equal to V_{Csn}, and the voltage across the diode D is $V_D = V_o - V_S$ during this mode. At the end of this mode, the voltage across the switch V_S increases to V_o, and therefore at $t = t_3$, $V_D = 0$.

2. Mode IV: $[t_3$ to $t_4]$: In this mode, the switch S and the diode D_{s1} remain ON and OFF, respectively. The diode D is turned ON due to the resonant circuit $(L_{sa} : L_{sb} : C_{sn})$. The input current is $I_{in} = I_D + I_{Ds1}$. Again, I_{Ds1} is given in equation (18.5):

$$I_{Ds1} = I_{Csn} + I_{Lsb} \tag{18.5}$$

The currents I_{Ls2} and i_{Cs} are given in equations (18.6) and (18.7):

$$i_{Lsb}(t) = \frac{-L_{s1}}{L_{sa} + L_{sb}}$$
$$\times\left(I_{in} - i_{Lsb}(t_3)\cos\omega_2(t - t_3) + \frac{1}{L_{sa} + L_{s2}}(L_{sa}I_{in} + L_{sb}i_{Lsb}(t_3))\right) \tag{18.6}$$

$$i_{Csn}(t) = i_{Csn}(t_3)\cos\omega_2(t - t_3) \tag{18.7}$$

where $\omega_2 = \dfrac{1}{\sqrt{\dfrac{L_{sa}L_{sb}}{L_{sa} + L_{sb}}}}$

The voltage V_{Csn} is increased by the charging current I_{Csn} from V_0, and it is given in equation (18.8):

$$V_s(t) - V_o = z_i i_{Csn}(t_3) \sin \omega_2 (t - t_3) \tag{18.8}$$

where $Z_2 = \dfrac{L_{sa} L_{sb}}{(L_{sa} + L_{sb}) C_{sn}}$.

The voltage across V_{sn} reaches its maximum value at the end of this mode (t_4), whereas the current i_{s2} will be $i_{L_{sb}}(t_4)$. The maximum voltage across the switch is given in equation (18.9):

$$V_{Smax} = V_0 + z_{2^i Csn}(t_3) \tag{18.9}$$

3. Mode V: [$t_4 - t_5$]: During this mode, the switch S remains OFF and the diodes D and $Ds1$ remain in the ON state. The input current is $I_{in} = I_D + I_{Ds1}$. The capacitor C_{sn} discharges through L_s and reaches $V_{Csn}(t_0)$. The current through the inductor I_{Lsb} reaches $i_{Lsb}(t_0)$ at the end of this mode.

18.2.2 DESIGN OF THE SBC

The diode D is turned OFF and it can be achieved from Ref. [12] and is given by equation (18.10):

$$L_{S1} > \frac{V_0 t_r}{I_{in}} \tag{18.10}$$

where t_r is the rise time of the switch. The reverse recovery current for the snubber diode Ds should reach zero at the end of the switching period t_5. The duration t_5 of mode V is given in equation (18.11):

$$\Delta t_5 = \frac{1}{\omega_2} \tan^{-1} \frac{Z_2}{Z_1 \tan(\omega_1 \Delta t_1)} \tag{18.11}$$

Based on equations (18.4) and (18.9)–(18.11), the parameters of the converter are designed.

The converter is designed for an input voltage of 24 V and an output voltage of 60 V. The switching frequency of the converter is taken as 100 kHz. The output power of the converter is 100 W, and the duty ratio D_T of the switch S is 0.6. The converter parameters are listed in Table 18.1.

18.3 EXPERIMENTAL RESULTS

The SBC and conventional converter BC are designed with the specifications as given in Table 18.1, and prototypes are developed for both the converters for a power rating of 100 W. The part number of the components and the cost of the components used in the prototype board of the SBC are given in Table 18.2.

TABLE 18.1
Parameters of the Converter

S. No	Parameters	Values	
1	Main inductor (L)	100	μH
2	Capacitor (C_0)	3.5 μ	F
3	Snubber inductors (L_{s1}; L_{s2})	2.5	μH
4	Snubber capacitor (C_s)	320	nf

TABLE 18.2
Part Number of the Components Used in Hardware

S. No	Parameters	Part No.	No. of Components	Cost (INR)
1	Main inductor (L)	7447070 Wurth Elektronik	1	250
2	Capacitor (Co)	Electrolytic	1	15
3	Snubber inductors (L_{sa}, L_{sb})	RCH895NP-2R5M	2	100
4	Snubber capacitor (C_{sn})	Electrolytic	1	15
5	Diode	FSQS05A065	2	120
6	MOSFET	IRLIZ14GPBF	1	100
7	Flood LED light	White light		1,000
	Overhead			350
	Total cost			1,950

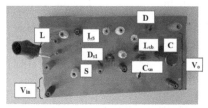

(a) Prototype board

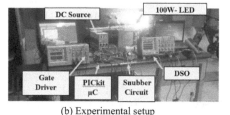

(b) Experimental setup

FIGURE 18.4 Prototype and experimental setup of a converter with a passive snubber.

The prototype board and the experimental setup for SBC are shown in Figure 18.4. The converter is experimented with an input voltage of 24 V. The switch S is operated with a ratio of 0.6. It is observed from Figure 18.5a that the input and output currents are 4.7 and 1.73 A, whereas the output voltage is 61.3 V. The voltage and current waveforms across the switch S are shown in Figure 18.5b. It is observed that ZCS using the passive snubber is achieved in the SBC.

The input and output voltages and currents are measured for different duty ratios for both the converters. The efficiency of the converters is calculated and shown in

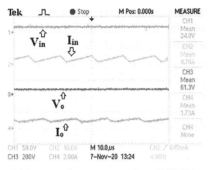

(a) Input and output waveform S (ZCS).

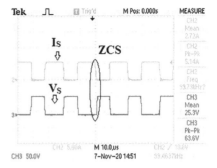

(b) Voltage and current through the switch.

FIGURE 18.5   Output waveforms for the SBC.

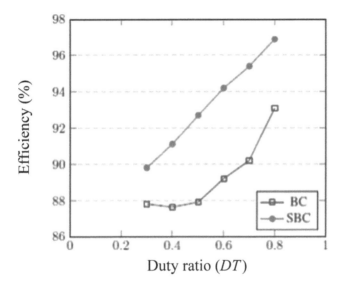

FIGURE 18.6   Efficiency vs. duty ratio of converters.

Figure 18.6. It is observed that the efficiency of the SBC is higher by 2% than the BC. Moreover, the luminous flux of the SBC is measured using lux −101 A. It is 12,500 lux for the SBC and 12,200 lux for the BC.

18.4   CASE STUDY OF LIGHTING DESIGN FOR A COMMERCIAL BUILDING

A tennis court of class 3 is considered for lighting design in a commercial building. Lighting is important for a tennis court as it increases the visibility and thus attracts participants and spectators to be a part of the game. A good lighting design [13] is required for the right amount of illumination because variations cause distraction. The general requirements for indoor tennis court lighting are its cost, maintenance, power supply, and the effect of natural lights. The existing lighting system in the court uses HPS lamps. The number of LED lights required for the tennis court involves a design procedure, and it is discussed in the following section.

18.4.1 Lighting Design for a Tennis Court

A two-pitch, indoor tennis court of 25 m length and 10 m width is considered for the lighting design. The factors that are required for design are horizontal illuminance (HI), uniformity of illuminance (UI), and glare (G). HI is the amount of light falling on the court surface and should be more than 300 lux. UI is the spread light over the surface and should be more than 0.5, whereas G is the information about light that impairs vision and should be less than 55 for indoor courts.

Metal halide, HPS, and LED lamps are some of the lights used in the indoor tennis court, of which LED lamps provide good color rendering and temperature. All these factors are considered along with the Energy Conservation Building Code (ECBC) [14] for lighting design. The lumen distribution for the tennis court is calculated as given in equation (18.12):

$$RI = \frac{lb}{h(l+b)} \tag{18.12}$$

where RI is the room index and is 0.79. The average lumens (E_{av}) is given in equation (18.13):

$$E_{av} = \frac{E_1 + E_2 + \ldots + E_n}{n} \tag{18.13}$$

E_{av} is 200 lm, and it is calculated by taking eight to nine different measuring points in the tennis court. The total available lumens and the load efficacy are given in equations (18.14) and (18.15):

$$\text{Total available lumens} \ (\varphi) = E_{av} * \text{Area of the room} \tag{18.14}$$

$$\text{Installed load efficacy} = \frac{\text{Total lumens on measurement plane}}{\text{Circuit watts}} \tag{18.15}$$

The φ and load efficiency for the tennis court from equations (18.14) and (18.15) are 50,000 lm and 125 lm/W, respectively. Similarly,

$$\text{Installed load efficiency ratio(ILER)} = \frac{\text{Installed load efficacy}}{\text{Target installed load efficacy}} \tag{18.16}$$

The installed load efficacy ratio is given in equation (18.16), and it is 3.47 for the tennis court.

Target installed load efficacy (TILE) for the sports application is 0.36 as per Ref. [14]. Based on the design, four LED lights are required to cover the entire tennis court of area 250 m², which gives the lux value of 50,000 lm. The parameters that are calculated are listed in Table 18.3.

The model of the indoor tennis court is simulated using DIALux Evo 9.1. For the simulation, a MGD-D100 LED lamp of 100 W is used, which has luminaire luminous flux of 12,520 lm, luminous efficacy of 114.1 lm/w, and light loss factor of 0.80 (fixed). The MGD-D100 is considered to be the equivalent of SBC LED driver. Apart from the factors given from equations (18.13)–(18.16), reflection factors of ceiling, walls, and floor are also taken into account. They are 70.0%, 0.0%, and 65.2%, respectively. The mounting height of the lights is taken as 2.9 m for the lighting design. The polar intensity diagram is shown in Figure 18.7 for LED lamps and HPS lamps.

The two lines of the light source, i.e., horizontal cone (blue) and vertical cone (red) as in Figure 18.7a, give detailed information about the spread of light in the

---

TABLE 18.3

Computed Requirements for Lighting Design

Parameters	Results	Unit
Room type	Indoor sports	-
Activity type	Tennis	-
Number of lamps	4	-
Length	25	M
Breadth/width	10	M
Number of illuminance points taken	8	
Average room illuminance	50,000	Lux
Measured/estimated circuit power	400	W
Installed lighting efficacy	125	lm/W
Target lighting efficacy	36	lm/W
Installed lighting efficacy ratio	3.47	p.u.

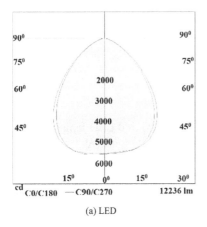
(a) LED

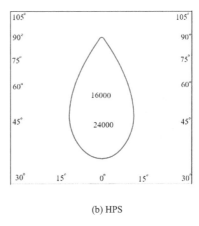
(b) HPS

FIGURE 18.7 Polar intensity diagram of LED and HPS.

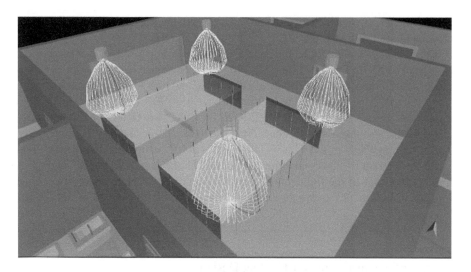

FIGURE 18.8   Lighting design in DIALux with LED lights.

2D graphical form. The horizontal angle covers 90°–270°, which is the side view of light distribution plane, whereas the vertical angle covers 0°–180°, which is the front view of the light distribution plane. The proposed LED light outspreads 4,000 lm at 45° and 5,500–6,000 lm at 30° with the even lighting system. On the other hand, the HPS lamp has very limited flux distribution around the source and it is shown in Figure 18.7b. The HPS lamp has the horizontal and vertical angle cover of 30° with 16 K lm. Moreover, the flux distribution is little uneven, which results in heavy intensity of light collating in closer place and dull intensity in the farther places. The DIALux simulation diagram for the tennis court is given in Figure 18.8.

The designed flux distribution of the tennis court is shown in Figure 18.9. It is observed that the intensity of the flux decrease radically and falls below the standard average illuminance (i.e., 200 lm). Therefore, to avoid irregular distribution, the sum of the illumination caused by the vertical and horizontal lights in the court is taken into account. It is found to be greater than 200 lm.

18.4.2 ENERGY SAVINGS IN THE TENNIS COURT

The energy consumed by the tennis court for the HPS and LED lamps is estimated in this section. The lamp efficacy and average output lumens (AVL) are greater for the LED lamp than for the HPS lamp, and they are estimated from Refs. [15,16]. Consider the illumination is required in the tennis court as 8 hours/day. The total energy consumed per year by HPS and LED lamps is given in Table 18.4.

From Table 18.4, it is observed that the energy consumed by LED lamps is 1,168 kWh, whereas the total energy consumed by the HPS lamps is 1,900 kWh. The installation cost of HPS and LED lamps is INR 15,000 and INR 30,000, respectively.

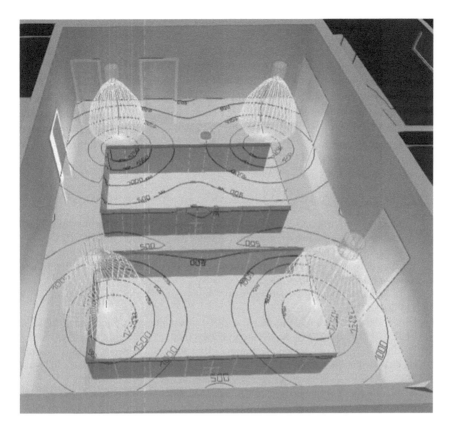

FIGURE 18.9 Flux distribution of LED lamps in a tennis court.

TABLE 18.4
Energy and Amount Saved by LED Lights

Parameters	LED	HPS
No. of lamps	4	2
Total load (W)	400	650
Energy consumed per annum per light (kWh)	292	950
Total energy consumed (kWh)	1,168	1,900
Total electricity bill (INR)	5,840	9,500

The amount saved per year in energy is INR 3,660. The payback period for HPS and LED lamps is 4 and 8.2 years, respectively. Though the capital cost of LEDs is higher than that of the HPS, LED lights have longer lifetime and the payback period is good in terms of energy savings. The LED lights continue to serve for a period of 10 years, and the amount saved during this period is INR 6,558.

18.5 CONCLUSION

A SBC and a conventional BC are designed and experimented for 100 W. The efficiency of the SBC is greater by 2%, whereas the output lumens is 12,500 lm. A new indoor lighting design is proposed, and it is simulated using DIALux software for LED (equivalent for SBD) and HPS lamps for a tennis court. It is evident that LED has good light spread than HPS. The energy consumed by both the lights is estimated per year, and for LED lights, it is lesser by 732 kWh than HPS. The LED lights continue to serve for a period of 10 years. Therefore, INR 3,660 is saved through energy savings using LED lights and the SBC.

REFERENCES

[1] F. Şener Yılmaz, Human factors in retail lighting design: an experimental subjective evaluation for sales areas. *Architectural Science Review* 61(3):156–170, 2018.

[2] I. Acosta, C. Varela, J. F. Molina, J. Navarro, and J. J. Sendra, Energy efficiency and lighting design in courtyards and atriums: a predictive method for daylight factors. *Applied Energy* 211:1216–1228, 2018.

[3] P. T. Tsankov, Lighting technologies, In *The Sun and Photovoltaic Technologies*, pp. 213–270, Springer, 2020.

[4] IEA, India 2020 Energy Policy Review, 2020. https://niti.gov.in/sites/default/files/2020-01/IEA-India%202020-In-depth-EnergyPolicy_0.pdf

[5] L. Mackay, N. H. van der Blij, L. Ramirez-Elizondo, and P. Bauer, Toward the universal dc distribution system. *Electric Power Components and Systems* 45(10):1032–1042, 2017.

[6] A. Marahatta, Y. Rajbhandari, A. Shrestha, A. Singh, A. Gachhadar, and A. Thapa, Priority-based low voltage dc microgrid system for rural electrification. *Energy Reports* 7:43–51, 2021.

[7] A. Chub, D. Vinnikov, F. Blaabjerg, and F. Z. Peng, A review of galvanically isolated impedance-source dc–dc converters. *IEEE Transactions on Power Electronics* 31(4):2808–2828, 2015.

[8] R. Srimathi and S. Hemamalini, Led boost driver topologies for low voltage dc distribution systems in smart structured buildings. *Electric Power Components and Systems* 46(10):1134–1146, 2018.

[9] R. Srimathi and S. Hemamalini, Performance analysis of single-stage led buck driver topologies for low-voltage dc distribution systems. *IETE Journal of Research* 1–13, 2019.

[10] T. Shamsi, M. Delshad, E. Adib, and M. R. Yazdani, A new simple-structure passive lossless snubber for dc–dc boost converters. *IEEE Transactions on Industrial Electronics* 68(3):2207–2214, 2020.

[11] J.-J. Yun, H.-J. Choe, Y.-H. Hwang, Y.-K. Park, and B. Kang, Improvement of power-conversion efficiency of a dc–dc boost converter using a passive snubber circuit. *IEEE Transactions on Industrial Electronics* 59(4):1808–1814, 2011.

[12] M. R. Amini and H. Farzanehfard, Novel family of pwm soft-single-switched dc–dc converters with coupled inductors. *IEEE Transactions on Industrial Electronics* 56(6):2108–2114, 2009.

[13] R. Srimathi, K. Eswaran, S. Hemamalini, and V. K. Kannan, Lighting design for a campus hostel building using single-stage efficient buck led driver. In *Advances in Smart Grid Technology*, pp. 337–351, Springer, 2020.

[14] Energy Conservation Building Code User Guide, 2011. https://beeindia.gov.in/sites/default/files/ECBC%20User%20Guide%20V-0.2%20%28Public%29.pdf

[15] IES, Hps to Led Conversion, 2011. https://www1.eere.energy.gov/buildings/publications/pdfs/ssl/silsby_msslc-phoenix2013.pdf

[16] A. Ikuzwe, X. Ye, and X. Xia, Energy-maintenance optimization for retrofitted lighting system incorporating luminous flux degradation to enhance visual comfort. *Applied Energy* 261:114379, 2020.

19 Implementation of Smart Grids and Case Studies through ETAP in Commercial and Official Buildings

*A.M. Chithralegha, N. Srinivasan,
and R. Balakrishnan*
L&T Construction

CONTENTS

19.1 ABOUT ETAP SOFTWARE

ETAP is the most thorough examination software for the plan and testing of electrical networks accessible well before execution [1–3].

The ETAP Load Flow Analysis module ascertains the transport voltages, branch power variables, flows, and force streams all through the electrical framework. ETAP

TABLE 19.1

Case Study Details 1

Sl. No	Study Case	EB GRID	DG Standby	Renewable Source
1	Case 1 – EB only	ON	OFF	OFF
2	Case 2 – EB and renewable source	ON	OFF	ON
3	Case 3 – DG only	OFF	ON	OFF
4	Case 4 – DG and renewable source	OFF	ON	ON

TABLE 19.2

Case Study Details 2

Sl. No	Source Details
I	Dry-type transformer – $1 \times 2,000$ kVA
	Diesel generator – $1 \times 1,750$ kVA
II	Renewable source details
	WTG – 10 kW
	WTG – 10 kW
	WTG – 40 kW
	Solar PV array – 251 kWp
	Solar PV array – 157.2 kWp

takes into consideration swing, voltage-managed, and unregulated force sources with various force matrices and generator associations. It is equipped for performing investigation on both radial and loop systems.

19.2 CASE STUDY – DETAIL

An 11-kV system grid study feeding an IT building loads is studied on a load flow study with renewable sources such as wind turbine and solar panels. A four-case study of priority is performed to understand the impact of power quality in terms of load and power factor. The load flow study of the system is elaborated in the following scenarios. Power source of voltage class for this facility distribution is an 11-kV supply. Case study details are given in Tables 19.1 and 19.2. Loads of lighting, raw power, UPS, chiller, AHUs, and data center are the lower-end distribution.

19.3 LOAD DETAILS OF THE IT OFFICE BUILDING CAMPUS

Table 19.3 indicates the load details.

Figure 19.1 indicates the single-line diagram (SLD) in ETAP with renewable resources (wind turbine and solar photovoltaic (PV) array).

Refer to Figure 19.1 for an overall SLD of distribution with solar panel and wind turbine.

TABLE 19.3
Load Details

I	Load Details	
A	Chiller 1	385 kW
b	Chiller 2	385 kW
c	Admin building	194.4 kVA
d	Canteen building	30 kVA
e	Research building	94.4 kVA
f	Emergency loads	17.78 kVA
g	Security building	43.98 kVA
h	UPS load	166.7 kVA
i	Capacitor bank – 600 kVAR	600 kVAR
j	Chiller 3	180 kVA
k	Research center	94.4 kVA
l	Emergency load	17.78 kVA
Grand total		1282 kVA

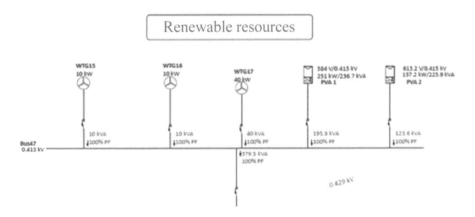

FIGURE 19.1 Single-line diagram in ETAP with renewable resources.

19.4 WIND TURBINE CHARACTERISTIC DESIGN IN ETAP 20.0.4 (WTG)

Figure 19.2 indicates the wind turbine characteristics.

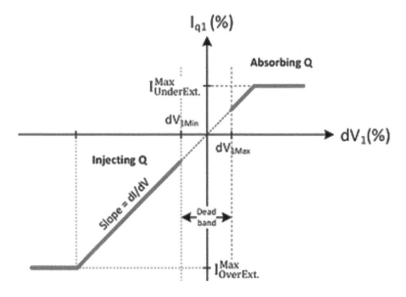

FIGURE 19.2   Wind turbine characteristics.

TABLE 19.4
Control and Operation Mode

Type	Control	Operation Mode
Type 1	WECC/UDM	Induction generator
Type 2	WECC/UDM	Induction generator
Type 3	WECC/UDM	Voltage control
Type 3	Generic/existing model (ETSP 5.0 onward)	Mvar control
Type 4	WECC/UDM	Induction generator

19.4.1   CONTROL MODE USED FOR WESTERN ELECTRIC COORDINATION COUNCIL (WECC) TYPE 4

The operation mode for WTG is fed as input. The operation mode for different types and controls is shown in selection criteria. Type 4, an advanced model, is scheduled to feed through the WECC control model. Table 19.4 indicates the same.

19.4.2   WIND TURBINE GENERATOR'S V_{MIN} AND V_{MAX} PARAMETERS

Figure 19.3 indicates the wind turbine's voltage parameters.

Reactive Power Capabilities
1. Types 1 and 2: Capability may be set to hold a fixed PF
2. Types 3 and 4: Typical range is 0.95 (capacitive) to 0.9 (inductive)
3. Type 5: The range is similar to that of a conventional synchronous machine.

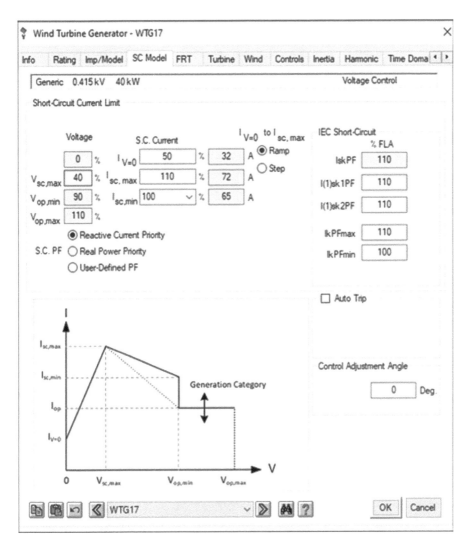

FIGURE 19.3 Wind turbine design parameters.

Note: Types 3, 4, and 5 may be able to provide VAR support even while not producing watts.

Types 3 and 4: These types are similar to a STATCOM.
Type 5: These types are similar to a synchronous condenser.

19.4.3 WIND TURBINE RAMP START AND STOP DETAILS

Figure 19.4 indicates the wind turbine ramp details.

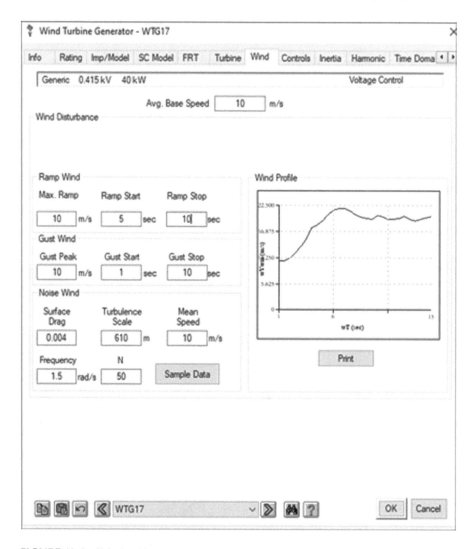

FIGURE 19.4   Wind turbine ramp details.

19.5   SOLAR PV ARRAY (PVA) CHARACTERISTIC DESIGN IN ETAP 20.0.4

Figure 19.5 indicates the PV characteristics. Figure 19.6 indicates that the current versus voltage (I-V) qualities of the PV module can be characterized in daylight and under dim conditions. In the main quadrant, the upper left of the I-V bend at zero voltage is known as the short out current. This is the current estimated with the yield terminals shorted (zero voltage). The base right of the bend at zero current is known as the open-circuit voltage. This is the voltage estimated with the yield terminals open (zero current).

Nonetheless, past a specific negative voltage, the intersection separates as in a diode, and the current ascents to a high worth. In obscurity, the current is zero for voltage up to the breakdown voltage which is equivalent to in the enlightened condition.

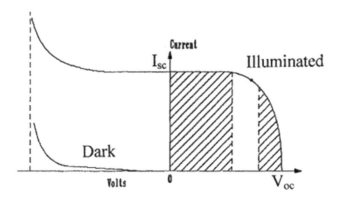

FIGURE 19.5 PV characteristics.

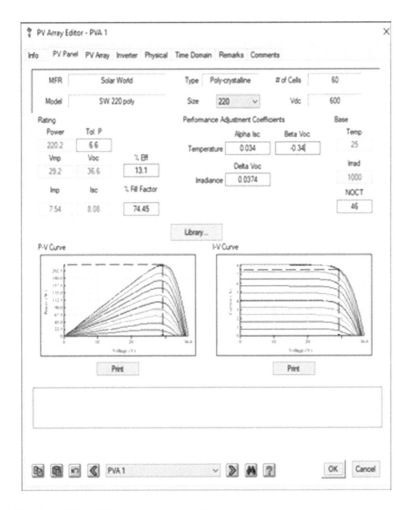

FIGURE 19.6 PV and IV panel characteristics.

Reliability assessment is a vital parameter in the life cycle of equipment. For example, in the initial design phase, the reliability assessment can **assess whether the system dependability meets the design requirement in the approval stage**, so that product improvement can be designed at the time of proposal of design itself. Figure 19.7 indicates the reliability parameters of PV panel characteristics.

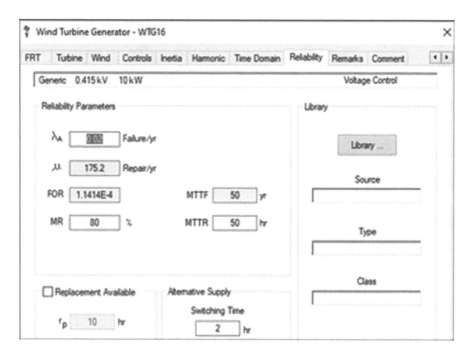

FIGURE 19.7   Reliability parameters of PV panel characteristics.

19.6 PVA DESIGN PARAMETERS

Figure 19.8 indicates the PVA design with reference to an inverter.

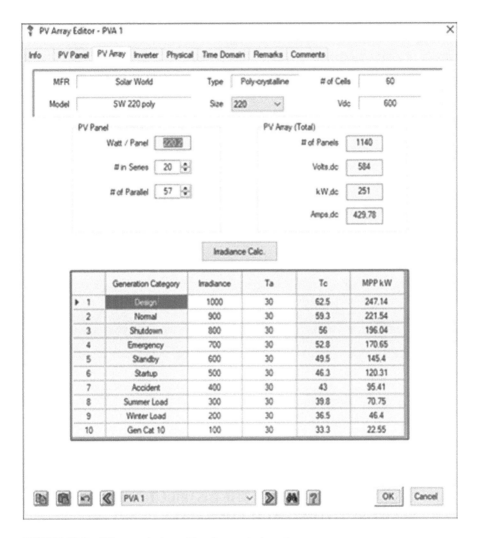

FIGURE 19.8 PV array design with reference to inverter.

19.6.1 INVERTER DETAILS OF SOLAR PVA

Figure 19.9 indicates the PVA design with reference to an inverter.

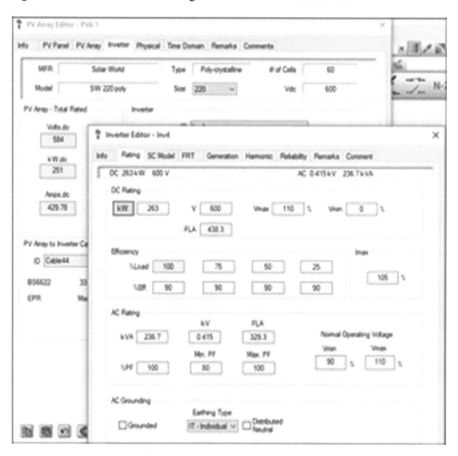

FIGURE 19.9   PV array design with reference to inverter.

19.7   RENEWABLE SOURCES IN THE DISTRIBUTION NETWORK FEEDING AT IT BUS 0.415 KV

Three wind turbines and two solar panels from different locations of the IT building office campus are used as in Figure 19.10.

19.7.1 OVERALL SLD IN ETAP 20.0.4

Figure 19.11 indicates the overall SLD of distribution with solar panel and wind turbine.

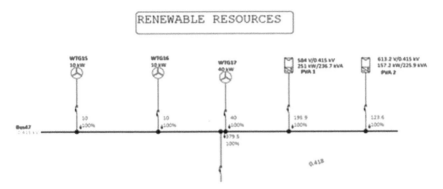

FIGURE 19.10 Renewable sources.

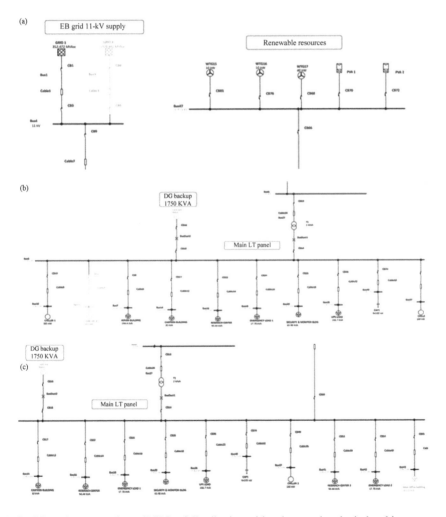

FIGURE 19.11 (a–c) Overall SLD of distribution with solar panel and wind turbine.

19.7.2 ANALYSIS IN DETAIL

Case 1 EB grid is shown in Figures 19.12 and 19.13.

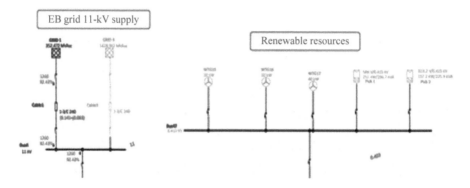

FIGURE 19.12 Case 1.

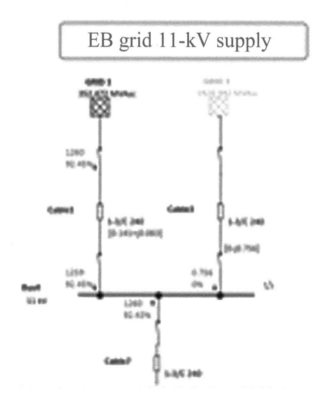

FIGURE 19.13 Case 1.

Case 2 EB grid with renewable source feeding loads is shown in Figures 19.14 and 19.15.

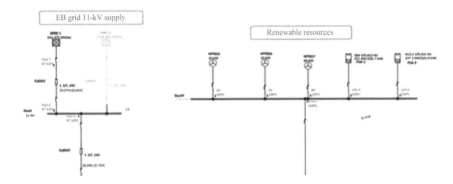

FIGURE 19.14   Case 2.

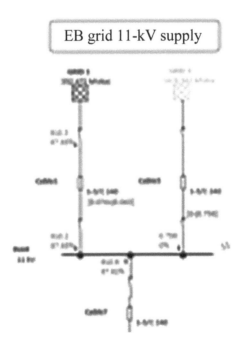

FIGURE 19.15   Case 2.

Case 3 DG backup feeding loads are shown in Figures 19.16 and 19.17.

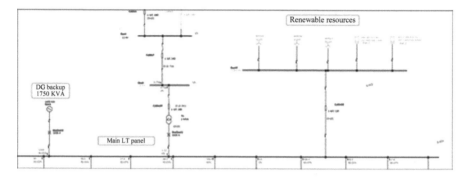

FIGURE 19.16   Case 3.

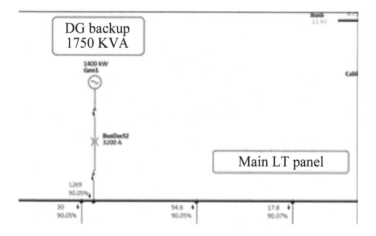

FIGURE 19.17   Case 3.

Case 4 EB grid with renewable source feeding loads is shown in Figures 19.18 and 19.19.

19.8   CONCLUSION

From different case scenarios of adding renewable energy source either with grid or with a standby backup such as diesel generator, we infer the results in Tables 19.5 and 19.6 on load flow analysis from the latest version of ETAP of 20.0.4. These results portray the energy savings on using renewable sources such as wind and solar inside an IT campus. Also, the impact of power factor using renewable sources on EB and DG operated system. Advancement in focus of improving the power factor should be considered while DER is being implemented in the EB or DG sources.

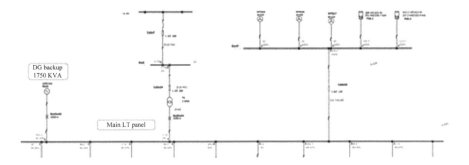

FIGURE 19.18   Case 4.

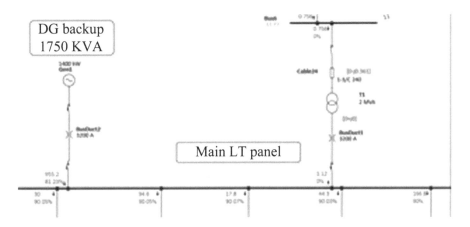

FIGURE 19.19   Case 4.

TABLE 19.5

Distributed Energy Resource (DER) with EB

Sl. No.	Study Case	EB Grid	DG Standby	Renewable Source	Result from ETAP	Energy Savings in %
1	Case 1 – EB	ON	OFF	OFF	1,260 kVA and 92.45% PF	27.73% savings in power on using DER
2	Case 2 – EB with renewable source	ON	OFF	ON	910.6 kVA and 87.65% PF	

The achievement of a power market can be surveyed by various quantifiable standards. The market should guarantee satisfactory capital assumption to keep up the unwavering quality and security of supply. With a lot of wind- and solar-oriented force being introduced, there is the necessity that this venture is in plant and innovation that supplement wind power qualities.

TABLE 19.6
Distributed Energy Resource (DER) with DG

Sl. No.	Study Case	EB Grid	DG Standby	Renewable Source	Result from ETAP	Energy Savings in %
3	Case 3 – DG	OFF	ON	OFF	1269 kVA and 90.05% PF	24.73% savings in power on using DER
4	Case 4 – DG with renewable source	OFF	ON	ON	955.1 kVA & 81.23% PF	

A further benefit of utilizing power conditioning hardware is that it maintains a strategic distance from the need to draw responsive force from the power organization. This advantages the utility and may reserve funds for the wind turbine administrator as receptive force is frequently dependent upon a charge. A further advantage increments if the wind turbine can sell energy with a leading power factor to the utility at a premium.

ABBREVIATIONS

BIM   Building Information Modeling
CFD   Computational Fluid Dynamic
DER   Distributed Energy Resource
DG   Diesel Generator
EB   Electricity Board
ETAP   Electrical Transient Analyzer Program
HVAC   Heat, Ventilation, and Air-Conditioning
SBEMS   Smart Building Energy Management Systems

REFERENCES

[1] www.etap.com – for details on wind turbine and solar PV array and its characteristics.
[2] Load flow Module stimulated for case study through ETAP 20.0.4 (Latest Version).
[3] Case study details of Electrical distribution loads with transformer, DG and Renewable sources received from an IT reputed firm on Request.

20 Data Logger-Aided Stand-Alone PV System for Rural Electrification

Aashiq A, Haniya Ashraf, Supraja Sivaviji, and O.V. Gnana Swathika
Vellore Institute of Technology

CONTENTS

ABBREVIATIONS

ADC	Analog-to-Digital Converter
DAQ	Data Acquisition
GUI	Graphical User Interface
IoT	Internet of Things
Isc	Short-Circuit Current
LDR	Liquid Detecting Resistor
PLC	Power Line Communication
PV	Photovoltaic
RES	Renewable Energy System
MCU	Microcontroller Unit
MPP	Maximum Power Point
MPPT	Maximum Power Point Tracker
RTC	Real-Time Clock
SAPV	Stand-Alone Photovoltaic
SVD	Solar Variability Data Logger
USART	Universal Synchronous and Asynchronous Serial Receiver and Transmitter
Voc	Open-Circuit Voltage

DOI: 10.1201/9781003240853-20

20.1   INTRODUCTION

With the advancement of modern technology and engineering, the demand of electricity in urban and industrial areas is huge. So, there exists a shortage of power supply facilities for rural households and remote locations. In such cases, solar energy is a promising solution to meet this electricity demand for rural areas in developing countries such as India [1]. In recent years, the fast evolution of renewable energies, mainly solar photovoltaic (PV) technology, has led to an explosion of PV installations throughout the world. PV systems generate electricity using energy from the sun.

PV systems can be categorized into two types: grid-connected PV systems and stand-alone PV (SAPV) systems.

Grid-connected PV systems are connected to the utility grid. SAPV systems are not connected to the grid, and theoretically, it provides electrical energy throughout the day, even at night when the sun is not available. SAPV system installations are limited by size and cost factors.

In this chapter, we focus on SAPV systems as these help in rural electrification since most of these are commonly located in remote locations in developing countries: Large programs have installed thousands of SAPV systems in rural areas of Asia, Africa, and Latin America [3]. SAPV systems are ideal for remote areas where an electrical source of power is either unavailable or impractical. However, for these small PV systems to provide good performance consistently, continuous monitoring and stability are necessary. In order to analyze the system and modify it for cost reduction and efficiency, a data capturing unit called data logger should be constructed that can store PV system data.

A data logger is a device that records data for a particular time, in relation to location – either with an available instrument or sensor, or through external instruments and sensors. They are usually small, battery-powered, and reliable. The common features for developing a general data logger are a microcontroller, internal memory for data storage (SD card), real-time clock (RTC) to time-stamp data, and various sensors (current, voltage, light, etc.) to collect real-time data. Some data loggers interface with a personal computer and use software to activate it and to view and analyze the collected data, while many others have a local interface device (such as keypad and liquid crystal display (LCD)) and can be used as a stand-alone device [2].

Data loggers have the capacity to automatically collect data on a 24-hour basis where data can be very accurately recorded for a particular hour, minute, or millisecond of a day, week, month, or year [2]. Upon activation, they are used to measure and record information for a specific duration of the monitoring period. This allows for a comprehensive and accurate setup of the electrical quantities and environmental conditions being monitored, such as solar radiation, air temperature, and humidity [2].

For efficient functioning and analysis of SAPV systems, the measurement of real-time PV parameters is absolutely necessary.

A. *Electrical parameters:* The number of parameters to be measured determines the number of inputs, and hence, it is important in data logger design [3]. Some electrical parameters of solar panels are maximum power point (MPP), voltage (V_{mpp}), current (I_{mpp}), open-circuit voltage (V_{oc}), and short-circuit current (I_{sc}).

B. *Meteorological parameters:* Weather-related parameters such as solar radiation, ambient temperature, relative humidity, cloud cover, wind speed, solar irradiance (sunlight intensity or power), clearness index, and sunshine duration are meteorological parameters. These too affect PV panel efficiency and hence must be recorded and analyzed.

C. *Data storage:* Data storage is an important process for carrying out any monitoring or performance analysis. The SAPV system was specifically designed for being installed in developing countries, and sending data may not always be possible due to the lack of coverage in rural areas. Hence, the method of data storage selected should be reliable – ensuring safety of collected data and work autonomously, requiring less maintenance [3].

Though the fast evolution of solar energy technologies has led to the installation of many systems all over the world for the past few years, most of these technologies have not yet achieved full development [4]. There is still a margin for improvement in cost and technology used. In the case of PV systems, detailed knowledge of the meteorological data (based on location) where the system will be installed and proper monitoring of the PV system performance are needed.

The limitations of data loggers in terms of unnecessary hardware led to the design and development of a new data logger in Ref. [4]. It is capable of monitoring PV systems at low cost with a flexible design. It uses a module of two high common-mode voltage difference amplifiers to measure the current from the PV modules by means of shunts and an AC transducer to measure the current from the micro-inverter [4]. The data logger developed in this study costs approximately 60 €, with eight analog inputs of 18 bits with a wide voltage range (up to 96 V), three analog inputs for cheap temperature sensors, and an unlimited number of digital temperature sensors [4].

Reference [5] introduces an economic data logger design to evaluate the performance of a stand-alone solar PV system. It uses a PIC microcontroller. An LDR that senses irradiance is used to assess the efficiency of the solar panel. The PIC microcontroller processes real-time values of current, voltage, and irradiation. Such data is then transmitted to a smartphone via GSM and then stored in an SD card as a text file.

Reference [6] describes an overall data logger system architecture that is capable of storing data from a few input channels and has broad memory space, which can store a large volume of analog signal data over an extended period of time, so maintenance is not needed. An estimated calculation is done to determine the capacity of a 4 GB SD card. One SD card was found to be capable of accommodating up to 1,294 days of data, i.e., approximately 3 years with a sampling time of 1 second, using four analog sensors. Hence, using two SD cards, the data logger can log data for 6 years without supervision [6].

Since installation cost is still high for renewable energy systems (RESs), their design optimization is desirable. However, such an effort requires a detailed knowledge of meteorological data of the location where the system will be installed. Reference [7] presents a computer-based data acquisition system based on Campbell's scientific data logger CR3000 that is optimally designed for RES monitoring and controlling PV parameters.

The data logger designed in Ref. [8] does logging, visualization, storage, and fundamental analysis of data. ThingSpeak – a free online platform for Internet of things (IoT) – is used to visualize the data. MATLAB is used to analyze the operations of the remote system. Arduino and Python programming languages are both used for data transmission. This data logger uses radiofrequency for weather data transmission to eliminate cable use and hence is cheap. It also includes an alarm feature that sends "connection failure" messages to the user's Twitter account [8].

The sensor-based microcontroller data logger system in Ref. [9] was developed for real-time monitoring of the PV system. The electrical and meteorological sensor outputs are first conditioned according to microcontroller analog-to-digital converter (ADC) input needs using precision electronic amplifier and active filter circuits and then digitized and processed using 32-bit ARM Cortex-M4 core microcontroller [9]. The established system records the data on an XLS file directly, although it is capable of using a USB port. Device output data and operational requirements for the environment are monitored. The system is efficient, effective, and inexpensive.

Nowadays, residential consumption of electricity has been increasing continuously. The residential PV system is supposed to give maximum performance to minimize the electricity cost. Power line communication (PLC) is a communication technology that allows for communication over existing power lines and does not need any extra communication lines.

Reference [10] proposes a user-friendly PV monitoring system based on a low-cost PLC. For cost reduction, the PLC module is developed without an expensive communication modem, which is done by using the ASK modulation scheme directly on the microcontroller unit (MCU) [10]. A smart app shows the aggregated data graphically so users can figure out the status of the whole PV system.

20.2   METHODOLOGIES

The rapid growth of PV systems has triggered the idea to create a low-cost monitoring system involving different types of integration. Currently, remote monitoring has gained popularity. Data loggers can be constructed in various ways by using different technologies. Each variation in technology and components used is to improve the overall efficiency and to lower the cost of the data loggers compared to commercial expensive ones in the market. Every data logger has its own limitations. A data logger cannot have all desirable features without a compromise in some other features. For example, a data logger with a lot of components such as higher number of I/O pins, though provides flexibility, can lead to an increase in cost. Too many features mean higher cost. Therefore, it is necessary to find an optimum and cost-effective solution based on location and application.

Different microcontrollers offer different features with varying costs and sizes. Hence, it is important to choose a microcontroller that suits the application accordingly.

The commonly used Arduino UNO uses an Atmel ATmega328 that allows for design flexibility and shorter program code. It is commonly used for writing command code in data loggers. It can send and receive data in both analog and digital forms. It consumes very low energy and can be programmed in many different languages. It can also change the processing or operation by modifying the program within the memory. The serial peripheral interface (SPI) of Atmel EEPROM can work at high clock speed and has faster memory writing time. This means that the high sampling rate allows for smooth logging of data. ATmega328P and ATmega328 are the same in every sense architecturally, except that ATmega328P consumes lesser power than ATmega328.

An Arduino UNO (microcontroller board based on ATmega328p) prototype of DC energy power logger in Ref. [11] measures and logs PV array DC and voltage, and calculates the generated energy in kilowatt-hour. The logger records every minute, which can be easily accessed by the user via MS Excel. An integrated LCD shows the real-time values of data. This prototype can only measure from 0 to 50 V and up to 50 A using a specific high-accuracy Hall effect IC module from the PV array [11].

Reference [12] implements the design of a stand-alone data logger that has RTC and SD card installed. ATmega328p microcontroller that is compatible with Arduino IDE is used. After reading values from sensors, RTC saves data to SD card in CSV format, which is compatible with MS Excel and is user-friendly. Data is also displayed on the LCD. The software is developed using Arduino IDE. This system is much cheaper compared to commercial devices in the market, but it is only suitable for measuring single PV panels in low range of DC voltage and current (0–30 V, 0–20 A) and not for large systems.

A low-cost data acquisition system is designed in Ref. [12] for improving and monitoring a PV system's electrical quantities, battery temperatures, and state of charge of battery using an ATmega328p microcontroller. This system is used along with a wall power data acquisition system for recording the regional power outages into micro-SD cards. The acquired data from both systems is used to monitor the status of the PV system and the local power grid. Very little addition of auxiliary sensors is required for this system.

Another high-performance and low-power usage microcontroller is the ATmega 2560 with low cost. The important elements that this microcontroller has are a stable ADC and universal synchronous and asynchronous serial receiver and transmitter (USART). The difference between Arduino UNO and ATmega2560 is that Arduino UNO is best suited for beginners who have just started using microcontrollers, while the Mega board is for enthusiasts who require lots of I/O pins for their projects – which are why it makes for a popular choice over other boards. ATmega2560 contributes incredibly in electronics control applications such as data logging and data acquisition systems as it is popularly known for its economic value and substantial flexibility among circuit designers [13].

The data logger developed in Ref. [13] is based on an Arduino Mega 2560 board along with ATmega2560 chip. It can store bulk data from input channels in large memory storage. For monitoring required parameters, a 240 W PV system is used where electrical parameters are tapped into the input channels of the data logger. The system converts the acquired raw data to digital input for data acquisition and stores it in an SD card. It is also equipped with a DS1307 RTC chip for data stamping in the SD card for every occurrence of logging. The performance of said data logger is synchronized with an existing data logger in the market, DataTaker DT80, and it was able to get almost ±0.5% accuracy by comparing the obtained data.

A low-cost Arduino Mega2560 microcontroller-based data acquisition system is built using halogen light in Ref. [14]. Halogen light ensures uniform dispersion of light. The intensity of light is changed using a dimmer regulator. This system is a complete solar emulator-DAQ system that can emulate the sunlight and monitor the panel parameters. It provides real-time updates of PV characteristics at any time of the year without relying on weather changes. The control interface is developed with the help of Arduino-MATLAB Target Package software.

Reference [15] developed a design of an Arduino-based data logger with IEC standard for a PV system. The data logger is used with the solar cell systems (fixed PV module and tracking PV module). The analog sensor signal is received by the Arduino microcontroller, which the processor board evaluates and displays on an LCD monitor at 20×4 (columns $\times$ rows) characters. The data logger in Ref. [15] was tested with the pyranometer, fixed PV module, and PV module monitoring. The values compared with the data from a regular data logger (Wisco DL 2100) showed the percentage values of the errors to be as follows: Maximum value of error from the recorded data was observed to be 25.6%, and the minimum was 0.

One type of data logger is a Bluetooth-based data logger. Bluetooth low-energy-based data loggers can transmit and measure the data wirelessly to Bluetooth-enabled mobile devices such as mobile phones over 10 m of distance range. These transmit data in the ISM band from 2.400 to 2.485 GHz range using UHF radio waves.

Reference [16] suggests a simple and cheap design for an Arduino UNO-based data logger. It is based on USB and Bluetooth and facilitates data transfer to a computer via SD card, USB port, or Bluetooth. The setup also uses HC-05 Bluetooth module and an SD card module.

USB-based data loggers are reusable, compact, and portable, and offer easy installation. It is a cheaper solution to data logging. Internal and external transducers can be used with the data logger in Ref. [16]. The internal sensor model monitors the process near the data logger location, while USB data loggers based on external sensors monitor the process far from the data logger location. Recorded data on USB data logger can be transferred to pen drive or by connecting the data logger to a computer via USB port [16].

The digital data logger circuit in Ref. [17] is built around the Microchip PIC18F4550 microcontroller to log solar energy parameters. A single printed circuit board (PCB) is used to fabricate the digital data logger. Since the accuracy of measurements and data are very important, the single-board computer designed for the implementation is capable of storing data every 3 seconds. The temperature-sensing device used is the LM35, which is an integrated circuit sensor that can be used to measure temperature with an electrical output proportional to the temperature (in °C) [17].

Reference [18] introduces a design for an ultra-low-power data logger for a SAPV energy system. Microcontroller ESP32-S2 is chosen in Ref. [18] as it has an ultra-low-power coprocessor. Using the deep-sleep mode of the microcontroller ESP32-S2 along with sensors reduced more than 90% self-power consumption of PV data logger. (Deep-sleep mode and Web server mode are two operating modes of the data logger that can be selected by a toggle switch.) Only the data stored in RTC memory can be processed during this mode. The power consumption of the board is 25 µA while only using RTC memory and RTC during deep-sleep mode. The data logger is programmed to wake up for 1 second during every 60 seconds of cycle, and the current consumption during wake-up time is 35 mA. ESP32-S2 collects data and stores it in an external memory card installed in a micro-SD card module. The ACS712 Hall Sensor is used to measure DC voltage from solar panels, and can measure a maximum of 20 A with a scale factor of 100 mV/A at a supply voltage of 5 V DC. The designed data logger consumes 64 Wh/year, which is quite low. Total cost of all the components used to design this data logger is $ 30 [18].

A monitoring and data logging system is developed in Ref. [19] by using a microcontroller and a single-board computer, and is suitable for monitoring small PV energy system's parameters with maximum current up to 25 A and voltages up to 55 V. This data logger was specifically designed for an off-grid solar PV unit installed in a private house. An STM32 microcontroller processes all the data. A Linux-based single-board computer receives the data from STM32 and transmits it to a remote server through the Internet every 5 minutes. The single-board computer can also store the data to its local SQL database in case of Internet connection failure [19].

In Ref. [20], an inexpensive prototype design and implementation of the microcontroller wireless connection-based DC logger are carried out for an existing 1.6 kWp single-phase grid-connected PV systems in Electrical Energy Conversion Research Laboratory (EECRL), Institute of Technology Bandung (ITB). The data logger developed by Arduino UNO microcontroller, telemetry 3D radio, and computer interface is designed to calculate the current, voltage, and power in real time. Delphi 7.0 software – which is a Windows-based Object Pascal development environment – is used to create a user-friendly smart application to show aggregated data graphically. Data retrieval can be done both online and offline.

Besides rural PV systems, data loggers are also used in other applications such as power plants and street lighting systems. These places also need monitoring systems to efficiently collect data for analysis and monitoring purposes. A few such applications are further researched for better understanding of technology used for stand-alone data loggers.

Reference [21] develops a low-cost and low-power-consuming data logger based on Atmel ATmega8 microcontroller, Dallas DS1307 Real-Time Clock, and 64 k Serial EEPROM from microchip, for an LED street lighting system based on PV solar cell in Jakarta Province, automatically, instead of an existing manual method. The data logger is portable and can be easily embedded inside the streetlight. It measures subsystem conditions such as PV solar cell, battery, charger/controller, LED lamp, and ambient temperature. It interfaces with a personal computer and uses software to activate the data logger for viewing and analysis of the collected data to check performance. The reports generated include daily energy used, total energy used, charging/discharging status, and failure condition. The power consumption is measured to be about 0.24 W (20 mA at 12 V) [21].

In large-scale PV power stations, a monitoring and control system is necessary to monitor and control system operation. A PV power station consists of PV array strings, storage battery bank, power conditioning unit, and electrical load appliances. In the operation of such large-sized stations (kilowatt or megawatt scales), the system performance should be monitored carefully and a proper decision must be taken in time [22].

The role of monitoring systems is to find issues and inform the operator by the type and location of failure to act accordingly. In Ref. [22], a monitoring and control system is proposed and implemented for PV power stations based on LabVIEW software and microcontroller interfacing. The microcontroller acts as a data logger. Laboratory Virtual Instrument Engineering Workbench (LabVIEW™) is a powerful, flexible instrumentation and analysis software application tool. LabVIEW™ has become an important tool in current emerging technologies and is widely adopted throughout academia, industry, and government laboratories as the standard for data acquisition, instrument control, and analysis [22].

The proposed system in Ref. [23] comprises four NI WSN 3202 sensor nodes and one WSN-9791 gateway module. The PC-LabVIEW-based station is proposed instead of NI cRIO-9014 real-time controller that is usually used in such applications – to reduce cost and limitations. The novelty of this system is that it can publish data over the Internet using LabVIEW Web server capabilities and deliver the operational performance data for the end-user location. User-friendly graphical user interface (GUI) enables the user to define and rearrange the monitored variables to suit the user's needs. In this way, the performance of the overall system in real rural conditions can be evaluated efficiently [23].

The newly designed monitoring system in Ref. [24], called the Solarmon-2.0, is built on special data collecting technologies. A special data logger, BBox, was installed at generating plants. It has secured communication that increases data protection. A higher frequency of data saving allows for higher accuracy of the mathematical models. Data is collected by data loggers and pushed periodically to the central server, where they may be accessible to the public or password protected with restricted access. This system was successfully tested on 65 PV arrays in the Czech Republic and several other countries. The monitoring system contributes to quality management of power plants and also provides data for scientific research. Even with a degradation of PV modules by 1% per year, there is a slight increase in the energy produced from PV sources in Central Europe according to the measurements seen. This increase is due to little precipitation and higher number of sun hours. Solarmon-2.0 software uses so-called responsive Web design where the system adapts to each display. Solarmon-2.0 software is available at app.solarmon.eu (user name: demo, password: demo1234) [24].

Most remote monitoring systems are too expensive and cable-connected where all the data are transmitted via fiber optics or local area network (LAN) connection [25]. In Ref. [25], a low-cost monitoring system for a small-scale stand-alone PV system is developed using a Raspberry Pi 3 microcontroller and Node-RED software.

Raspberry Pi is used as it is cheap and small, while its performance is almost comparable to a normal computer. The system in Ref. [25] is able to acquire, store, and display solar PV parameters in real time. It acts as a virtual monitoring system by displaying parameters using the GUI provided in Node.js. Node.js is an open platform built in JavaScript language and is suitable for real-time data monitoring purposes. Furthermore, JavaScript has the advantage of an outstanding event model, suitable for a parallel style of programming in which a node of works runs separately when executed. With this system, real-time data of the solar PV system can be monitored and analyzed remotely by using computer or mobile phone or any device that has an Internet connection by using localhost data provided, without interruption [25].

20.3   OPTIMIZATION AND ANALYSIS

To achieve an efficient output of a solar monitoring system, optimization and analysis techniques must be carried out. Research papers that predict the power losses along with data predictions with examples are discussed below. Data predictions help to identify the likelihood of future outcomes based on recorded past data. The goal is to go beyond the present knowledge to provide the best outcome of what will be required for the future. Possibly, fault conditions can also be predicted.

Various techniques can be used for future data analysis using the data logger. Reference [26] uses artificial neural networks to predict the energy output based on evolutionary programming artificial neural network (EPANN), which optimizes the training parameters of one hidden layer feedforward model.

Detection of solar intensity is possible to be measured and calculated using various parameters. Brightness is not a reliable factor to judge the true quality of available sunlight. Reference [27] highlights how the growing conditions (sunlight, temperature, moisture) and specific wavelength of light affect photosynthesis. A real-time LED-based light analyzer using data logger, which can measure the wavelength band of the sunlight and differentiate atmospheric conditions, is implemented in Ref. [27].

To calculate the maximum peak power, the data logger in Ref. [28] is integrated with current and voltage sensors to record parameters that affect the efficiency of the PV system. The data (power) is then compared with the calculated power, and maximum power point tracking (MPPT) method is used to collect efficient data of peak power. To compare the power in different areas, a data logger with MPPT method is used as a portable recorder [31]. As the temperature increases, the performance of crystalline silicon PV systems reduces. It was found that in Ref. [29], the nonuniform distribution of temperature and the thermal stress between the back and front surfaces of the PV panel were reduced just by introducing a water film cooling system. The efficiency of the overall system is increased by 15%, and power is improved by 32 W.

A programmable low-cost electrical system is used to compare the functions of PV modules and performance under variables (temperature, humidity). An Arduino collects the output parameters to calculate the predicted output power. The output observed in Ref. [30] was like that of the manufacturer's data list and the numerical simulation.

The energy yield of different modules is extracted under optimum condition by monitoring every individual module under MPP conditions. The system in Ref. [31] is used to monitor the long-term stability and performance of the models under realistic field conditions. The data measured is collected and transmitted to a multi-channel data logger by the means of analog voltages proportional to the current and voltage at the MPP of the modules. The suggested system in Ref. [31] is fast to track quick changes in power output and irradiance caused by the PV module.

A PV module as discussed in Ref. [32] was developed to obtain the output characteristics simulated by the model under normal operating conditions. The output characteristic graph is obtained using simulation tools such as PSpice and Solar.Pro. The VI and PV graphs are successfully recorded for future analysis.

One of the main challenges faced while incorporating local solar PV systems is the cost. To reduce the cost of a SAPV system, the battery capacity is reduced in Ref. [33] by analyzing the solar data of a particular place over a year by using a data logger. The paper concludes that the generated power is always higher than the power consumed by the loads in various seasons [33].

Another way to cut cost is discussed in Ref. [34] where a low-cost irradiance PV sensor is tested to ensure whether that could match the solar variability measurement of high-cost pyranometers. Then, a solar variability data logger (SVD) is developed to measure the solar variability (irradiance sensor, power, GPS, communication, and data logging). A comparative study is done on the collected data using a data logger from SVD to show the difference in solar variability by place and day [34].

To make the PV system work with maximum efficiency, Ref. [35] gives a solution. To detect repairs and to localize faults in a system, monitoring a PV system at the panel level comes in handy. The monitoring time between PV panels is synchronized to compare them under similar conditions. Reference [35] suggests a low-cost PLC module for simultaneous monitoring of PV panels. Threshold was set, and it was compared with mean voltage to detect faults. The proposed method is user-friendly and is an economic solution.

The efficiency of a solar panel is determined by environmental external factors such as temperature humidity, place, wind, and dust. The loss of power due to the accumulation of dust over a period of 55 days was measured in Ref. [36]. A 9% reduction in power generation was observed because of an unclean panel in Jaipur [36].

Reference [37] presents performance evaluation and capacity augmentation of a laboratory-scale PV array installation at Brunel University, UK, having a monocrystalline and a heterojunction with intrinsic thin-layer (HIT) PV array. Reports on the collection of online electrical and meteorological data and utilization of the data to evaluate the performance of the PV arrays under different solar irradiance conditions are given.

The design and strategy of installation of the various test bed components, their selection criteria, the characteristics, and the software strategy used with the test beds are discussed in Ref. [38]. Performance indicators such as performance ratio (P_R), final yields (Y_f), and reference yields (Y_r) are estimated. This information is very important for large PV installations. It explains installations of three PV technology-based test beds – single junction amorphous silicon (a-Si), HIT, and multi-crystalline silicon (c-Si) along with data logger and climate monitoring station. Total performance ratio of all three technologies was evaluated for 20-month duration. The PR rating of a-Si and HIT was found to be about 4% and 7% higher compared to c-Si modules [38].

20.4 IoT AND SMART TECHNOLOGIES

The IoT has a vision in which the Internet extends into the real world, to incorporate everyday objects in real life. It is achieved by usage of wireless networks. IoT allows objects to be sensed or controlled remotely over existing network infrastructure, so it creates opportunities for pure integration of the physical world into computer-based systems. It guarantees improved energy efficiency, accuracy, minimization of supervision time, and economic benefit in addition to reduced human intervention in solar PV monitoring systems.

IoT-based remote solar PV monitoring system ensures that the PV cells of solar panels are functioning perfectly by tracking the PV system's power production. It is easier to generate intelligent insights for faster and better analysis and decision-making as one gains real-time visibility into the amount of kilowatt-hours of electricity produced by the solar panels. Advantages include 24/7 access to real-time data and reports, increased availability of equipment, and reduced downtime and related O&M costs.

Reference [39] describes a plan that aims to develop a solar PV remote monitoring system, where an IoT-based cost-efficient solar PV monitoring system is designed using a GPRS module and a microcontroller that gives real-time analysis of the model at continuous intervals, which also further aids in fault detections if any. Measured data can be accessed from anywhere around the globe via the Internet.

Similarly, a smart-centric, user-friendly, remote monitoring solution as proposed in Ref. [40] uses ESP32 microcontroller-based data logger system to develop a pico solar home system in a rural area of a developing country. This system stores parameters in a micro-SD card and displays them on a Blynk App. Data can be downloaded directly from the Web page for analysis and verification. The hardware prototype uses only four sensors for humidity, voltage, temperature, and current sensing. An android app shows all real-time data for efficient monitoring and maintenance.

A cost-effective IoT-based solar PV system designed in Ref. [41] using open-source tools enables remote, real-time monitoring of an off-grid, stand-alone system via Web. The smart data logger records electrical and environmental parameters at regular intervals that are necessary to determine energy production. This information can be tracked using the Internet at any time from any place.

In Ref. [42], the authors have proposed an idea of monitoring essential parameters of a SAPV system such as voltage, current, and panel temperature with the help of IoT using ThingSpeak platform and sensors, following which power is computed. The parameters are transferred to cloud with help of node MCU ESP8266, which is programmed in C in Esplora. An android application is developed to fetch the data from the cloud.

Solar home systems fitted with an efficient MCU Arduino technology are illustrated in Ref. [43]. The proposed novel low-cost data logger can measure up to 14 electric and climatic parameters with the required accuracy established by the IEC61724 standard. Connectivity of the system via 3G and IoT allows for remote visualization of real-time data via Web or smartphone. This system fails only above 40°C, but can withstand all other harsh temperatures. Due to its stackable hardware design, several SAPV systems installed close enough can be monitored using a single data logger as a central node.

In contrast to small home systems, the use of Arduino can also be extended to large-scale models such as solar power stations as mentioned in Ref. [44]. In this IoT-based monitoring system, Arduino UNO and Arduino NANO act as the central brain of the system and ESP8266 is used as a communication module to deliver wireless connectivity via IEEE 802.11 communication protocol. Sensors are used to detect various PV parameters; a servomotor is used to implement the angle control technique to change direction of the solar array to get maximum output. Data from the cloud can be used for optimization. A disadvantage of this system is that ESP8266 has limited bandwidth, so it can only provide limited connectivity.

Digital control and data logging for solar power plants can also be done using SBC Raspberry Pi as proposed in Ref. [45]. In this model, a double pivot sun-oriented tracker effectively monitors the weather, tracks the sun, and changes its position accordingly to improve power yield. The solar panel direction is controlled by preset timings and associated degree of rotation. It consists of sensors, solar panels, two DC motors, and motor driver IC to drive motors for rotation. It can calculate wind speed and direction as well as moisture level and temperature. The values of parameters are displayed on ThingSpeak server, an IoT platform, which is stored on cloud. With this project, about 29% increase in yield is found [45].

Another idea as suggested in Ref. [46] describes a cost-effective, open-source IoT solution that collects and monitors real-time data intelligently. The solution is designed as a laboratory prototype that could be extended to monitor large-scale PV stations with minor adjustments. Five parameters, i.e., power, current, voltage, temperature, and irradiance level, are measured. Results of data collected in Cloud are visually represented. It sends an alert notification or an email to the user if monitored power goes below threshold value, i.e., when fault occurs.

For achieving maximum power and effective utilization of such systems, a method to monitor dust accumulated on solar panels is followed in Ref. [47]. It also displays malfunctioned solar panels and other real-time functioning data. IoT is incorporated to help in remote monitoring and sends data to cloud. Users can view current, average, and pass parameters via a GUI. A central controller alerts the user if values fall below specified conditions.

Such IoT-integrated control systems or units used for monitoring in Refs. [48, 49] make it easier to transmit data efficiently and help to predict fault in the PV system beforehand. The maintenance is also simple and cost-effective. Furthermore, in Ref. [48], modified daylight-based trackers with a cleaning system are implemented to remotely monitor a solar PV plant for performance evaluation. It facilitates maintenance, fault detection, and analysis of the plant based on past data in addition to real-time monitoring. Cleaning is done with the help of a DC servomotor and microcontroller.

Another use of wireless microcontroller technology is implemented in Ref. [50]. A customized, cost-efficient wireless data logger is designed, which can perform quickly and in wide and short ranges. It is based on IEEE 802.15.4 and GSM technology. To communicate with wireless sensor nodes in the designed data logger, the IEEE 802.15.4-based Zigbee JN5148 M003 wireless 32-bit microcontroller is used. Zigbee is generally applied in external sensor networks where there is no telephone or Internet coverage and there are many sensor nodes [50].

Similarly, in Ref. [51], a wireless remote monitoring and control system of a solar photovoltaic distributed generator (PV-DG) is designed to supply energy to microgrids. A 1.28 kWp setup of PV polycrystalline modules generates DC voltage, which is converted to AC voltage with the aid of an inverter, and then connected to a single-phase utility grid. A wireless sensor network (WSN)-integrated control system is utilized. The wireless communication technology incorporates a full-duplex digital system using the Zigbee protocol, based on the IEEE 802.15.4 standard for wireless personal area network (WPAN). The remote monitoring control system is executed using a digital signal processor (DSP), and human–machine interface (HMI) software is used to interact with remote sensor systems (RSSs).

With electricity costs continuing to rise and energy resources becoming scarce, it is certainly important for people to have awareness on utilizing energy in the most efficient way. IoT has seen a great increase in advancements lately, thus presenting significant potential for changing behavior and providing services that are efficient. Smart technologies such as IoT have found a way for homes and businesses to do this, therefore promoting energy conservation to a large extent.

20.5 CONCLUSION

While main sources of energy such as fossil fuels, gas, and nuclear energy may not be replaced entirely yet, the futuristic and more practical goal is to make electricity via solar energy available for all, especially in rural areas in developing countries that do not have access to the grid for electricity. SAPV systems open up the possibility to achieve rural electrification. Solar energy is the most trending, clean alternative as it is abundant and environment-friendly. The only problem is the cost of harnessing it, i.e., the initial cost of setup and maintenance. But with technological advancements, the cost of devices is decreasing rapidly. Hence, a good efficient monitoring system is what is required, which can perform major tasks automatically and can provide data to the user whenever and wherever needed. Data loggers for SAPV systems come into important play here, as a cost-effective and versatile device that can monitor and store data automatically without human intervention. Various ways of constructing a data logger are discussed in this chapter. Different technologies for implementing a data logger based on the application and location are discussed, and their methods are analyzed. With the boom of IoT to cope up with rapidly changing technology, IoT also proves to be an efficient smart solution for monitoring solar installations via the Internet (cloud). Some of the best IoT technologies are highlighted. This chapter also explores numerous optimization and analysis techniques to further increase the efficiency of a SAPV system. Hence, an extensive research is concluded and the potential of rural electrification through solar power systems is assessed.

REFERENCES

[1] Shuvo, A., Rahman, M., Nahian, A., and Himel, M. (2019). Design & implementation of a low-cost data logger for solar home system. *International Journal of Engineering and Management Research* 9(1), 32–20.

[2] Tripathi, S. K., Ojha, P., Singh, K. A., and Baliyan, A. K. (2017). Solar data logger. *IJSTE - International Journal of Science Technology & Engineering* 3(9).

[3] Lopez-Vargas, A., Fuentes, M., Garcia, M., and Munoz-Rodriguez, F. (2019). Low-cost datalogger intended for remote monitoring of solar photovoltaic standalone systems based on Arduino. *IEEE Sensors Journal* 19(11), 4308–4320.

[4] Fuentes, M., Vivar, M., Burgos, J., Aguilera, J., and Vacas, J. (2014). Design of an accurate, low-cost autonomous data logger for PV system monitoring using Arduino that complies with IEC standards. *Solar Energy Materials and Solar Cells* 130, 529–543.

[5] Malagond, G., Bagewadi, G., Pooja, L., Ronad, B.F., and Jangamshetti, S. (2015). Design and development of data logger for standalone SPV systems. *Int J Emerg Technol Adv Eng* 5(5).

[6] Mahzan, N., Omar, A., Mohammad Noor, S., and Mohd Rodzi, M. (2013). Design of data logger with multiple SD cards. *2013 IEEE Conference on Clean Energy and Technology (CEAT)*, pp. 175–180.

[7] Engin, M., (2017). Open source embedded data logger design for PV system monitoring. *2017 6th Mediterranean Conference on Embedded Computing (MECO)*, pp. 1–5.

[8] Jiang, B., and Iqbal, M. T. (2019). Open-source data logging and data visualization for an isolated PV system. *MDPI Journals* 8, 424.

[9] Nehovski, N., Tomchev, N., Djamiykov, T., and Asparuhova, K. (2018). Data logger for small solar systems. *2018 IX National Conference with International Participation (ELECTRONICA)*, pp. 619–623.

[10] Han, J., Lee, I., and Kim, S. (2015). User-friendly monitoring system for residential PV system based on low-cost power line communication. *IEEE Transactions on Consumer Electronics* 61(2), 175–180.

[11] Ruzaimi, A., Shafie, S., Hassan, W., Azis, N., Yaacob, M., and Supeni, E. (2019). Microcontroller based DC energy logger for off-grid PV system application. *2019 IEEE International Circuits and Systems Symposium (ICSyS)*, pp. 1–5.

[12] Hadi, M. S., Afandi, A. N., Wibawa, A. P., Ahmar, A. S., and Saputra, K. H. (2018). Stand-alone data logger for solar panel energy system with RTC and SD card. *Journal of Physics: Conference Series* 1028, 012065.

[13] Fanourakis, S., Wang, K., McCarthy, P., and Jiao, L. (2017). Low-cost data acquisition systems for photovoltaic system monitoring and usage statistics. *IOP Conference Series: Earth and Environmental Science* 93, 012048.

[14] Mahzan, N. N., Omar, A. M., Rimon, L., Noor, S. Z. M., and Rosselan, M. Z. (2017). Design and development of an arduino based data logger for photovoltaic monitoring system. *International Journal of Simulation: Systems, Science & Technology* 17, 15.1–15.5.

[15] Ahmed, O., Sayed, H., Jalal, K., Mahmood, D., and Habeeb, W. (2019). Design and implementation of an indoor solar emulator based low-cost autonomous data logger for PV system monitoring. *International Journal of Power Electronics and Drive Systems (IJPEDS)* 10(3), 1645.

[16] Rewthong, O., Boonbumroong, U., Mamee, T., Eamthanakul, B., Luewarasirikul, N., and Tabkit, N. (2019). Design of the data logger with IEC standard for PV system. *Proceedings of the 8th International Conference on Informatics, Environment, Energy and Applications - IEEA'19*, pp. 234–237.

[17] Singh, T., and Thakur, R. (2019). Design and development of PV solar panel data logger. *International Journal of Computer Sciences and Engineering* 7(4), 364–369.

[18] Akposionu, K. N. (2012). Design and fabrication of a low-cost data logger for solar energy parameters. *Journal of Energy Technologies and Policy*, 2.

[19] Rehman, A., and Iqbal, M. (2020). Design of an ultra-low powered data-logger for standalone PV energy systems. *European Journal of Electrical Engineering and Computer Science*, 4(6).

[20] Krishna, A. M., Rao, K. P., Prakash, M. B., and Ramchander, N. (2012). Data acquisition system for performance monitoring of solar photovoltaic (PV) power generation. *International Journal of Engineering Research & Technology (IJERT)*, 1(7).

[21] Iskandar, H. R., Purwadi, A., Rizqiawan, A., and Heryana, N. (2016). Prototype Development of a Low Cost Data Logger and Monitoring System for PV Application. *2016 3rd Conference on Power Engineering and Renewable Energy (ICPERE)*, pp. 171–177.

[22] Purwadi, A., Haroen, Y., Ali, F. Y., Heryana, N., Nurafiat, D., and Assegaf, A. (2011). Prototype development of a low cost data logger for PV based LED street lighting system. *Proceedings of the 2011 International Conference on Electrical Engineering and Informatics*, pp. 1–5.

[23] Zahran, M., Atia, Y., Alhosseen, A., and El-Sayed, I. (2010). Wired and wireless remote control of PV system. *WSEAS Transactions on Systems and Control* 5, 656–666.

[24] Beránek, V., Olšan, T., Libra, M., Poulek, V., Sedláček, J., Dang, M., and Tyukhov, I. (2018). New monitoring system for photovoltaic power plants' management. *Energies* 11(10), 2495.

[25] Othman, N., Zainodin, M., Anuar, N., and Damanhuri, N. (2017). Remote monitoring system development via raspberry-Pi for small scale standalone PV plant. *2017 7th IEEE International Conference on Control System, Computing and Engineering (ICCSCE)*, pp. 360–365.

[26] Henni, O., Belarbi, M., Haddouche, K., and Belarbi, E. H. (2017). Design and implementation of a low-cost characterization system for photovoltaic solar panels. *International Journal of Renewable Energy Research* 7, 1586–1594.

[27] Oates, M., Ruiz-Canales, A., Ferrández-Villena, M., and López, A. (2017). A low cost sunlight analyzer and data logger measuring radiation. *Computers and Electronics in Agriculture* 143, 38–48.

[28] Effendi, A., Dewi, A., and Ismail, F. (2018). Data logger development to evaluate potential area of solar energy. *MATEC Web of Conferences* 215, 01014.

[29] Mah, C., Lim, B., Wong, C., Tan, M., Chong, K., and Lai, A. (2019). Investigating the performance improvement of a photovoltaic system in a tropical climate using water cooling method. *Energy Procedia* 159, 78–83.

[30] Henni, O., Belarbi, M., Haddouche, K., and Belarbi, E. H. (2017). Design and implementation of a low-cost characterization system for photovoltaic solar panels. *International Journal of Renewable Energy Research* 7, 1586–1594.

[31] Zimmermann, U., and Edoff, M. (2012). A maximum power point tracker for long-term logging of PV module performance. *IEEE Journal of Photovoltaics* 2(1), 47–55.

[32] Tang, K., Chao, K., Chao, Y., and Chen, J. (2012). Design and implementation of a simulator for photovoltaic modules. *International Journal of Photoenergy* 2012, 1–6.

[33] Khan, S., Raihan, S., Habibullah, M., and Abrar, S. (2017). Reducing the cost of solar home system using the data from data logger. *2017 IEEE International Conference on Smart Grid and Smart Cities (ICSGSC)*, pp. 37–41.

[34] Lave, M., Stein, J., and Smith, R. (2016). Solar variability datalogger. *Journal of Solar Energy Engineering* 138(5).

[35] Han, J., Jeong, J., Lee, I., and Kim, S. (2017). Low-cost monitoring of photovoltaic systems at panel level in residential homes based on power line communication. *IEEE Transactions on Consumer Electronics* 63(4), 435–441.

[36] Gupta, V., Raj, P., and Yadav, A. (2017). Investigate the effect of dust deposition on the performance of solar PV module using LABVIEW based data logger. *2017 IEEE International Conference on Power, Control, Signals and Instrumentation Engineering (ICPCSI)*, pp. 742–747.

[37] Chowdhury, S., Day, P., Taylor, G., Chowdhury, S., Markvart, T., and Song, Y. (2008). Supervisory data acquisition and performance analysis of a PV array installation with data logger. *2008 IEEE Power and Energy Society General Meeting - Conversion and Delivery of Electrical Energy in the 21st Century*, pp. 1–8.

[38] Magare. D., Sastry, O. S., Gupta, R., Kumar, A., and Sinha, A. (2012). Data logging strategy of photovoltaic (PV) module test beds. *27th European Photovoltaic Solar Energy Conference*, pp. 3259–3262.

[39] Kekre, A., and Gawre, S. (2017). Solar photovoltaic remote monitoring system using IOT. *2017 International Conference on Recent Innovations in Signal processing and Embedded Systems (RISE)*, pp. 619–623.

[40] Tellawar, M. P., and Chamat, N. (2019). An IOT based smart solar photovoltaic remote monitoring system. *International Journal of Engineering Research & Technology (IJERT)* 8(9).

[41] Lopez-Vargas, A., Fuentes, M., and Vivar, M. (2018). On the application of IoT for real-time monitoring of small stand-alone PV systems: results from a new smart data logger. *2018 IEEE 7th World Conference on Photovoltaic Energy Conversion (WCPEC) (A Joint Conference of 45th IEEE PVSC, 28th PVSEC & 34th EU PVSEC)*, pp. 605–607.

[42] Sarswat, S., Yadav, I., and Maurya, S. K. (2019). Real time monitoring of solar PV parameters using IoT. *International Journal of Innovative Technology and Exploring Engineering (IJITEE)* 9(1S).

[43] Lopez-Vargas, A., Fuentes, M., and Vivar, M. (2019). IoT application for real-time monitoring of solar home systems based on Arduino with 3G connectivity. *IEEE Sensors Journal* 19(2), 679–691.

[44] Awais, S., Moeenuddin, S., Ibrahim, A. M., Ammara, S., and Bilal, F. (2020). IoT based solar power plant monitoring system. *International Journal of Advanced Science and Technology* 29(9s), 7668–7677.

[45] Kadam, A., Kasar, T., Sonje, S., & Tavse, S. (2015). Digital control and data logging for solar power plant using raspberry Pi. *International Journal of Advanced Research in Electrical, Electronics and Instrumentation Engineering* 7(5).

[46] Cheddadi, Y., Cheddadi, H., Cheddadi, F., Errahimi, F., and Es-sbai, N. (2020). Design and implementation of an intelligent low-cost IoT solution for energy monitoring of photovoltaic stations. *SN Applied Sciences* 2(7).

[47] Lokesh Babu, R. L. R., Rambabu, D., Rajesh Naidu, A., Prasad, R. D., and Gopi Krishna, P. (2018). IoT enabled solar power monitoring system. *International Journal of Engineering & Technology* 7(3), 526.

[48] Pitchaimuthu, P., and Sridhar, K. (2019). An IoT based smart solar photovoltaic remote monitoring and control system. *International Journal of Science & Engineering Development Research* 4(5), 354–358.

[49] Adhya, S., Saha, D., Das, A., Jana, J., and Saha, H. (2016). An IoT based smart solar photovoltaic remote monitoring and control unit. *2016 2nd International Conference on Control, Instrumentation, Energy & Communication (CIEC)*, pp. 432–436.

[50] Asif, M., Ali, M., Ahmad, N., Haq, S., Jan, T., and Arshad, M. (2016). Design and development of a data logger based on IEEE 802.15.4/ZigBee and GSM. *Proceedings of the Pakistan Academy of Sciences: Pakistan Academy of Sciences A. Physical and Computational Sciences*, 53(1), 37–48.

[51] Andreoni Lopez, M. E., Galdeano Mantiñan, F. J., and Molina, M. G. (2012). Implementation of wireless remote monitoring and control of solar photovoltaic (PV) system. *2012 Sixth IEEE/PES Transmission and Distribution: Latin America Conference and Exposition (T&D-LA)*, pp. 1–6.

21 Smart Solar Modules for Smart Buildings

Nasrin I. Shaikh
Nowrosjee Wadia College

R. Rajapriya and Milind Shrinivas Dangate
Vellore Institute of Technology

CONTENTS

21.1 INTRODUCTION

Directly converting sunlight to electricity is the most elegant form to convert energy through the photovoltaic (PV) effect. As the global community recognizes the limited resources and detrimental effects of conventional fossil energy, additional efforts will be made to utilize renewables. Trends indicate global energy demands will increase by 56%, with fossil fuels accounting for 80% of energy production, and increased CO_2 emissions of 46% by 2040 [5]. According to the US Energy Information Administration (EIA), future US electricity generation by non-hydropower renewables will nearly double by 2040 as indicated in Figure 21.1 [6].

Solar energy presents a viable alternative source of energy to meet the global energy and environmental demands. An increase in non-hydropower renewables is expected as new technologies demonstrate increased efficiency and reduced manufacturing costs. The depletion of easily accessible fossil fuel resources will drive energy costs upward and stimulate popular interest in sustainable energy technologies. As we

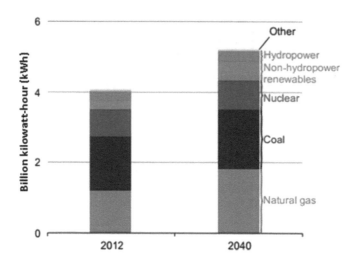

FIGURE 21.1   Electricity generation in the USA in total between 2012 and 2040 [6].

move toward a globally modernized and expanding population, supplying the global energy demands will continue to be a significant challenge. Figure 21.2 demonstrates the exponential energy consumption, even as population growth starts to stabilize. The need to balance current energy practices with renewable resources is critical for achieving a sustainable energy future.

A sustainable energy portfolio would include a blend of renewable technologies such as wind, hydro, bio, and solar in addition to limited fossil fuel and nuclear sources. Annual global installations of new PV expanded significantly from 2000 to 2011, as presented in Figure 21.3. The EIA predicts electricity generation from renewables to nearly double by 2040, with an increasing renewable generating capacity resulting from solar as seen in Figures 21.3 and 21.4, respectively. While it is accepted that most renewable technologies have energy production limitations due to variation in environmental conditions, the ability to offset peak demands is significant.

Solar PV technologies are well suited to be installed at point of use in residential and commercial spaces and are becoming common among dual-purpose shade structures and commercial rooftop installations. PVs are ideal for these applications due to the simplicity, versatility, reliability, and low environmental impact of installations. Large utility installations are becoming more popular in the arid southwest where the land is not suitable for agriculture, but has excellent solar resources [9]. With the development of the smart grid and advancements in the electrical distribution infrastructure, renewable energy integration will become increasingly manageable [10]. In addition to the improvements in grid transmission, integration of renewable technologies into construction has opened new venues previously too valuable for low-density energy production [11].

Capitalizing on the available solar resources would enable practical diversification of the global energy portfolio achieving grid parity, and making solar a sustainable industry without subsidies. While the program is for reductions in manufacturing costs, improvements to conversion efficiency, and public acceptance, PVs will

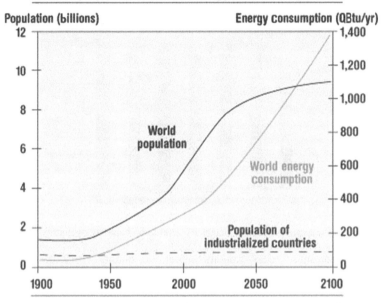

FIGURE 21.2 World population and energy demand growth [7].

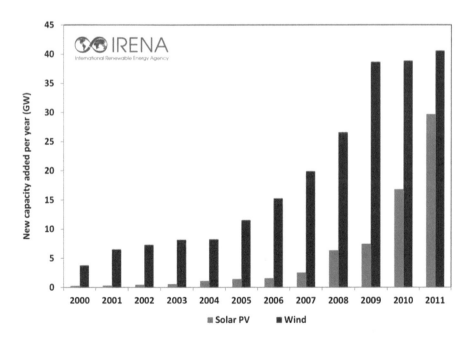

FIGURE 21.3 New installed capacity of wind and PV from 2000 to 2011 [8].

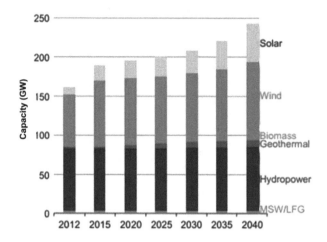

FIGURE 21.4   Predicted renewable electricity generation capacity by energy source [6].

become an important source of economical premium-quality power. Trends described that balance-of-system (BOS) cost and module cost will continue to diminish with increased capacity and improved technologies. The SunShot was launched by the Department of Energy (DOE), and an initiative was taken to stimulate the revitalization of PV research by the private sector after continued decreases in the market share of solar manufacturing. The SunShot program has a target cost of electricity from PV systems of $.02 per kilowatt-hour (kWh) by 2025; effectively years old, it has already accomplished a 60% reduction in the cost of utility-scale systems [12]. Market trends demonstrate the significant reduction in the cost per watt ($/watt) of module pricing as seen in Figure 21.5, which directly affects the residential and utility system costs.

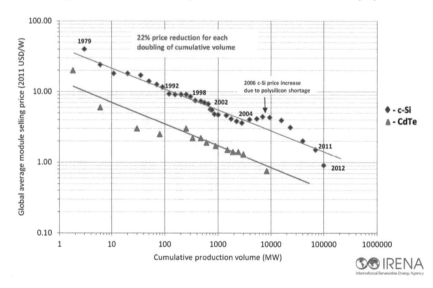

FIGURE 21.5   Price forecast curve of a solar system after 2008 [8].

21.2 FUNDAMENTALS OF SOLAR CELLS

The p-n diode was made usually by joining one or more semiconductor materials to form a solar cell. A solar cell converts light photons to electrical current through the PV effect. A significant impact on the energy picture will be observed even if very little amount of solar energy from the total amount reaching the earth surface will be transformed using solar PV. Sunlight is a portion of the electromagnetic radiation emitting from the sun. We define sunlight as a spectrum of discrete units of energy called photons, whose energy is based on wavelength. Potential efficiency of a solar cell getting affected by the light reaching the earth has a spectral distribution. At the atmospheric edge of the earth, the spectrum is considered as solar constant of air mass zero (AM0). As the radiation passes though earth's atmosphere, it interacts with gases and aerosols that cause significant losses due to absorption, reflection, and scattering. The spectrum incident on a surface at sea level with a zenith angle of 37° is denoted air mass 1.5 global (AM1.5g; Figure 21.6) and has a normalized intensity of 1,000 W/m² [13]. This distribution accounts for both diffuse and direct irradiance, as solar cells are effective at converting both. The AM1.5 spectral distribution is recognized as the standard for terrestrial characterization.

Only higher energy photons than that of the semiconductor's band gap are absorbed and contribute to the electrical energy generation. The difference in energy between the top of the valence band and the bottom of the conduction band is called band gap of a direct-gap semiconductor, defining the energy necessary to generate an electron–hole pair. Figure 21.7 represents the band diagram of a basic single-junction solar cell under illumination. The Shockley–Queisser model predicts the efficiency limits for the solar cells with p-n junction using, for example, an absorber layer. Graphically represented in Figure 21.8, one can see the ideal theoretical conversion efficiency of CdTe solar devices.

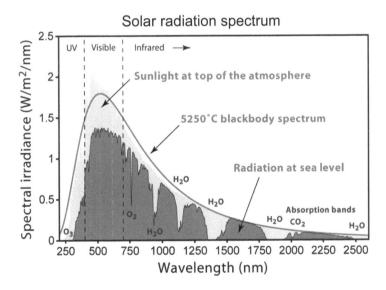

FIGURE 21.6 The AM1.5g solar radiation spectrum [13,14].

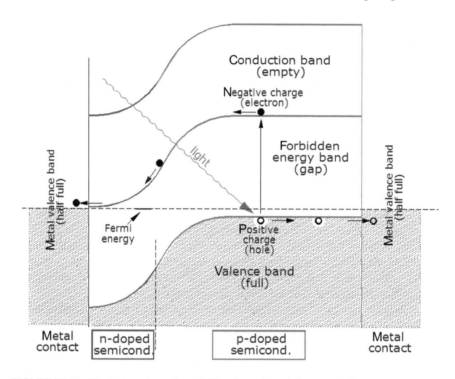

FIGURE 21.7   Single-junction solar cell illuminated band diagram [15].

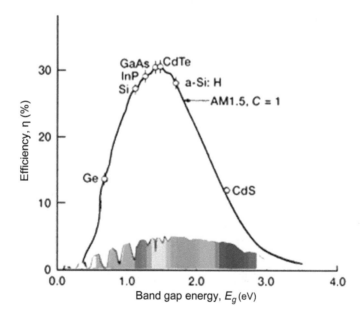

FIGURE 21.8   The Shockley–Queisser efficiency limit of single-junction solar cells [16,17].

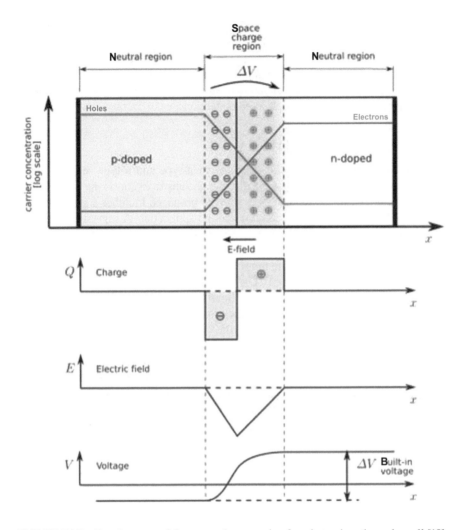

FIGURE 21.9 Development of the space charge region for a heterojunction solar cell [18].

A p-n heterojunction is established when different n-type and p-type semiconductors were joined. Demonstrated as in Figure 21.9, carriers immediately diffuse across the junction resulting ionized atoms and establishing neutral regions of fixed charge. This space charge region becomes depleted of both electrons and holes, producing a built-in field that acts as a barrier to the further drift of majority carriers and presents a low-resistance path to minority carriers. With illumination, the built-in field causes charge separation, driving electrons to the n-side and holes to the p-side inducing photogenerated current, and performing work when attached to an external load.

Heterojunctions are limited by the formation of interface states which work to trap charges at the junction, reducing the built-in voltage. A quantifiable performance gap exists between lab and ideal efficiencies, related to open-circuit voltage

(V_{oc}), fill factor (FF), and short-circuit current (J_{sc}). Performance losses due to these parameters are well explained by Sites and others [19,20]. Developing better correlations between process, performance, and structure should allow improvements to all parameters. The performance gap between lab and commercial efficiencies is expected to reduce as advanced manufacturing techniques are achieved to improve window layers, uniformity, and cell interconnections.

21.3 CdS/CdTe SOLAR CELLS

Solar cells are formed through the assembly of n-type and p-type semiconductor layers. Thin-film structures are formed directly on substrates consisting of polymers, glass, metal, or ceramics. Solar devices may be produced in either a superstrate or substrate configuration depending on process and application. A superstrate configuration is most common for fabrication ease and encapsulation quality. In this configuration, the incident light first passes through the substrate before being absorbed in the photojunction. This structure requires the use of a highly transparent substrate and transparent conductive oxide (TCO) layer.

Terrestrial solar applications have an optimal band gap of ~1.5 eV [17,21]. Selection of a material system for large-scale applications is based on potential device efficiency and manufacturing costs resulting the metric of \$/kWh. The use of CdTe is attractive as it is a direct-gap semiconductor with a band gap of 1.45 eV and a high absorption coefficient. These qualities result high conversion efficiencies of wavelengths up to 810 nm with absorber layers of 2 μm thickness. The typical heterojunction CdTe device utilizes an n-type CdS window layer with a band gap of 2.42 eV. This wide band gap allows transmission of wavelengths greater than 510 nm when thin layers are used.

Various techniques for the manufacture of CdS/CdTe thin-film solar cells are possible, including chemical bath deposition (CBD), closed-space sublimation (CSS), physical vapor deposition (PVD), and vapor transport deposition (VTD). Both CdS and CdTe are well suited to CSS. Films grown by CSS have many advantages over low-temperature processes, producing an excellent microstructure, and device performance when treated with a high-temperature CdCl2 process. High-volume manufacturing benefits of CSS include high deposition rates, large area uniformity, and excellent material utilization. Record device efficiencies have been produced by CSS and VTD techniques. The CdTe solar cell described herein is produced using a novel heated pocket deposition (HPD), a variation of CSS.

21.4 ROADMAP OF RESEARCH

There are many approaches currently under investigation for improving the efficiency of CdTe solar devices. The focus of this research work was on efficiency improvement and volume manufacturing techniques.

Multiple techniques for depositing the absorber layer, including the PVD co-sublimation source (Figure 21.10) developed at Colorado State University, Fort Collins, USA. Preliminary results demonstrate the feasibility of creating a higher band gap material by means of co-sublimation and are accordingly a focused effort. The roadmap

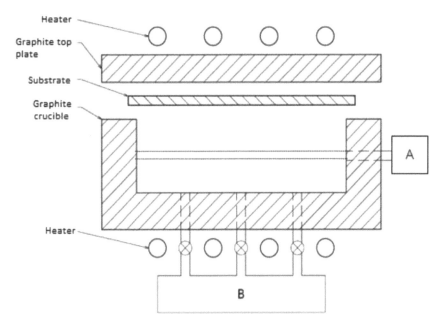

FIGURE 21.10 PVD co-sublimation source used for the device fabrication.

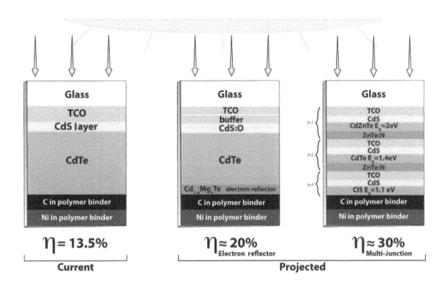

FIGURE 21.11 MEL and NSF I\UCRC research roadmap to high-efficiency photovoltaics.

of the Materials Engineering Laboratory (MEL) and National Science Foundation Next Generation Photovoltaics Industry/University Centre for Research Collaboration (I/UCRC) to pursue higher efficiency multiple-junction cells is presented in Figure 21.11. The devices in this study have the structure marked as current.

21.5 THE ADVANCED RESEARCH DEPOSITION SYSTEM

The original R&D system developed at the MEL was a continuous, in-line process, transporting samples from air to vacuum to air (AVA) on a continuous belt conveyor represented in Figure 21.12 [22]. The AVA system was conceived to demonstrate volume manufacturing of the thin-film solar process and eventually achieved success with the start-up of Abound Solar (formerly known as AVA Solar). The HPD process combined with a conveyor transport is suitable for production, but is limiting in all aspects of research [23,24]. The AVA system utilized a multistage molecular seal instead of common load-lock configurations. This necessitated specific substrate tolerances to maintain process pressure, limiting the ability to test alternative substrate geometries. There was no ability to generate partial or intermediate films while simultaneously producing full devices [25]. As the samples were continuously exposed to the high-temperature sources, motion had to be maintained. This meant that a large number of transient samples were produced between target set points, and the potential for process drift to occur was common.

To achieve experimental flexibility and maintain a concept capable of volume manufacturing, an in-line research tool was necessary. Considering current and future process limitations, a linear tool configuration utilizing the well-established HPD technique was developed. The ARDS design enabled hardware and process flexibility, with the potential for future additions. The linear layout was selected for ease of sample transport and necessitated by an existing boundary limitation [26]. The hardware was designed to be modular, providing a spatial envelope for which prototype hardware could be tested, including alternative deposition techniques. The single-sample transport mechanism allows variation of substrate type, process sequence, dwell times, and the ability to generate intermediate films for characterization. The ARDS provides a highly versatile platform for development of next-generation solar cells.

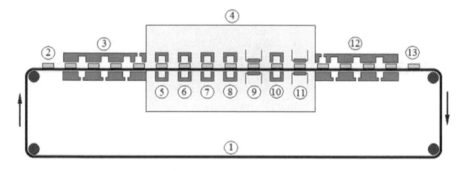

FIGURE 21.12 Diagram of the AVA system developed at the MEL. (1) Continuous conveyor belt, (2) base glass substrate, (3) seal, (4) vacuum chamber, (5) substrate heating, (6) CdS deposition, (7) CdTe deposition, (8) CdCl2 deposition, (9) CdCl2 annealing and stripping, (10) back contact formation, (11) back contact annealing, (12) molecular seal, and (13) completed device.

21.6 THE NEED FOR AUTOMATION OF EXPERIMENTS

Experimental efficiency minimizes time, resources, and waste. Carefully controlling the manufacturing process is necessary to produce meaningful and repeatable results. Automation of the ARDS is accomplished by the integration of process controls through a custom LabVIEW interface. Operators are able to define process recipes, load a substrate, and let the machine do the rest, eliminating user variation [27]. All valve actuation for transfer, pumping, and purging is automated, minimizing disruptions to the process environment and potential damage to the equipment. Compared to the previous AVA system, the ARDS executes experimentation with significantly fewer transient samples.

21.6.1 Methodology

The final assembly, start-up, characterization, and process development of the ARDS have been completed. Several modifications to the hardware needed to obtain reliable performance of the ARDS were completed in this study. These include (i) thermal shielding of sources, (ii) vapor shielding between the sources for chemical isolation, (iii) modification to the substrate top heaters for improved process control, and (iv) installation of pyrometer for monitoring substrate temperatures. The novel method of embedded Nichrome heaters has proven to be robust and enabled the development of a co-sublimation source, and the co-sublimation process has enabled the deposition of ternary alloys such as $Cd1-xMgxTe$ [28] at high throughput. Incorporation of this heating method into the source shutter has increased potential run length to more than 75 full samples between replenishing the source charge.

The effect of the process conditions on the properties of the films for each process has been quantified. A standard operating procedure (SOP) has been developed to produce ~12.7% efficient devices. As a result of this research, device efficiencies of $12\% \pm 0.5\%$ are repeatedly produced. Incorporation of 2.5% oxygen into the CdS:O window layer resulted in a device with J_{sc} of 25 mA/cm^2 and efficiency of 16.2%. A total of 110 runs were completed during this research, producing ~1,050 samples. To date, more than 565 runs have been completed, due to the efforts of this research and successful operation of the ARDS. The ARDS has been found to be highly effective to study the effect of process conditions on device performance. The capabilities of the ARDS are significant for new process development in CdTe PV and present many future opportunities.

The ARDS continues to operate with a high level of research efficiency and repeatability. A recent demonstration of ARDS capability on TEC 12D glass produced nine devices from a single sample with a 12.7% mean efficiency with organic back contacts dip-coated, and a library of organic compounds is presented in Table 21.1.

The ARDS has been instrumental in the demonstration of Cd1-xMgxTe (CMT) Electron Reflector (ER), resulting in V_{oc} and Time Resolved Photo-Luminescence (TRPL) improvements. Efforts to improve window layers employ the flexibility of the ARDS to produce completed devices utilizing CdS:O, indium tin oxide (ITO), and high-resistivity transparent (HRT) TCO buffer layers.

TABLE 21.1

Library of Organic Compounds

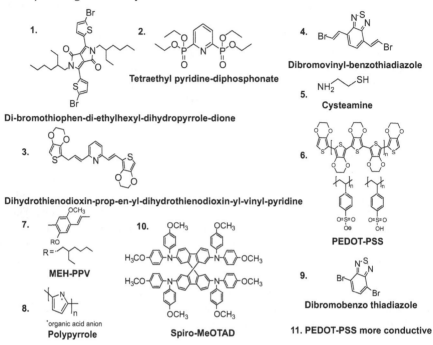

21.6.2 Results and Discussion

In this context, we have used soda-lime glass TEC 12D coated with TCO which was commercial. On this substrate, CdS and CdTe were deposited with the CSS process in an inert atmosphere using a single vacuum chamber. The above table represents the library of compounds we have deposited as a back contact in the device. Figure 21.13 shows detailed understanding of all parameters during the deposition process as well as order of different layers that we have deposited using an inert atmosphere Figure 21.14.

The CdTe material is proved to be a special material for establishing ohmic contact because of high work function of the material. When a metal contact is applied on cadmium telluride, this often results in the formation of the Schottky barrier. The barrier acts as a diode in the opposite direction, thus blocking the photogenerated charge carriers. This process is known as back barrier, or back diode, or back surface field and can affect all major photoabsorbers such as cadmium telluride, silicon, and CuInSe2. The nice layer of conductive small molecules and polymers is considered as an alternative for the traditional back contact layer in the conventional cell structure.

With all of these aspects in any solar cell device, back contact also requires to have a good conducting material for better efficiency. Keeping this in mind, we have deposited a very thin layer of hole-conducting small molecules and polymers onto

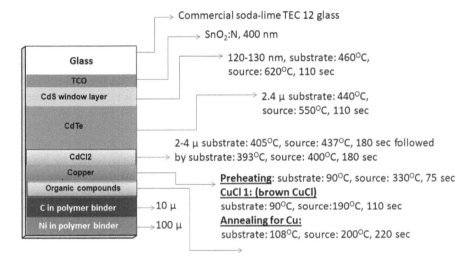

FIGURE 21.13 Fabricated device.

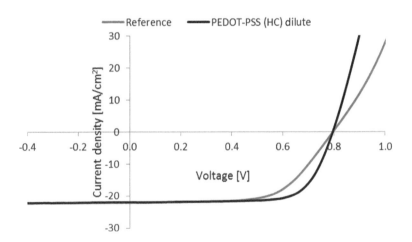

FIGURE 21.14 J-V characteristic of the complete structure under irradiance of 100 mW/cm² intensity.

the cadmium telluride layer by the dip-coating method. All these organic compounds proved to be good hole conductors in other devices, and for dip-coating, dispersion of chloroform or water was used. We have found that a clear solution of organic small molecules or polymers gives a good result.

21.6.3 CONCLUSION

21.6.3.1 Challenges Addressed

Many challenges were addressed during the development of the ARDS. Developing methods and methodologies to address operational and process control was a significant portion of the research performed. Achieving proper sample preparation and end effector cleaning eliminated front-end variation. Developing solutions to mechanical deficiencies and optimizing those that could not be altered resulted in a robust research system. Process development is a perpetual process and will evolve as a new discovery and research interest. The progress made during this research provided a strong platform for further improvements, and much is still to be learned from the samples generated as a result.

21.6.3.2 Perspective

The opportunity to bring the ARDS online was a unique challenge, testing mechanical, thermal, and vacuum deposition competencies. The process development required extensive material and device characterization, expanding our knowledge of key techniques and the interpretation thereof. Successful device characterization is accomplished only after developing a strong understanding of advanced physics principles and acceptance of the unique characteristics of CdTe PV.

21.7 FUTURE WORK

Currently, the ARDS is configured only with HPD sources for thin-film deposition. Future research of more advanced devices shown in will require alternative hardware such as sputtering, (metal organic chemical vapor deposition) MOCVD, and (physical vapor deposition) PVD, as well as longer cluster tools. The ARDS is well suited for the continued development of alternative substrate/TCO combinations, window layer optimization, PECSS, and CMT. The ARDS currently does not support in situ characterization. The incorporation of basic optical methods such as ellipsometry would benefit the investigation of film properties while under continuous vacuum.

ACKNOWLEDGMENT

Dr. Milind Dangate was supported by a Bhaskara Fellowship from the Department of Science and Technology, Government of India, for this research.

REFERENCES

[1] M. A. Green, K. E. Yoshihiro Hishikawa, W. Warta, and E. D. Dunlop, Solar cell efficiency tables (version46). *Progress in Photovoltaics* 23(7), 805–812, 2015.

[2] E. Wesoff, Exclusive: First Solar's CTO Discusses Record 18.6% Efficient Thin-Film Module. 2015.

[3] W. J. Beek, M. M. Wienk, and R. A. Janssen, Hybrid polymer solar cells based on zinc oxide. *Journal of Materials Chemistry* 15(29), 2985–2988, 2005.

[4] S. Bereznev, et al., Hybrid solar cells based on CuInS 2 and organic buffer–sensitizer layers. *Thin Solid Films* 515(15), 5759–5762, 2007.

[5] U.S. Energy Information Administration, "International Energy Outlook 2014", 2014.

[6] Administration, U.S. Energy Information, "Annual Energy Outlook 2014 with Projections to 2040", Administration, U.S. Energy Information, 2014.

[7] U.S. Department of Energy, "National Energy Technology Laboratory", [Online]. Available: http://energy.gov/eere/sunshot/about [Accessed 8 June 2014].

[8] Energy's tricky tradeoffs, *Science* 239, 786–787, 2010.

[9] S. Blumsack and A. Fernandez, Ready or not, here comes the smart grid! *Energy* 37, 1, 61–68, 2012.

[10] F. Trubiano, *Design and Construction of High-performance Homes: Building Envelopes, Renewable Energies and Integrated Practice*, Abingdon, Oxon, Routledge, 2013.

[11] IRENA, "Solar Photovoltaic", International Renewable Energy Agency, 2012. [Online]. Available: http://costing.irena.org/charts/solar-photovoltaic.aspx [Accessed 29 June 2014].

[12] Energy, U.S. Department of, "About", Office of Energy Efficiency & Renewable Energy, [Online]. Available: http://energy.gov/eere/sunshot/about. [Accessed 8 June 2014].

[13] NREL, "Reference Solar Spectral Irradiance: Air Mass 1.5", [Online]. Available: http://rredc.nrel.gov/solar/spectra/am1.5/ [Accessed 18 June 2014].

[14] J. L. Gray, The physics of the solar cell, in *Handbook of Photovoltaic Science and Engineering*, San Francisco, John Wiley and Sons, 2003, Chapter 3.

[15] Wikipedia, "Theory of Solar Cells", [Online]. Available: http://en.wikipedia.org/wiki/Theory_of_solar_cells [Accessed 12 June 2014].

[16] First Solar, "Thin Film Module Technology", First Solar, 2012. [Online]. Available: http://www.firstsolar.com/en/technologies-and-capabilities/pv-modules/first-solar-series-3-black-module/cdte-technology [Accessed 19 June 2014].

[17] W. Shockley and H. J. Queisser, Detailed balance limit of efficiency of p-n junction solar cells. *Journal of Applied Physics* 32, 510–519, 1961.

[18] Wikipedia, "Depletion Region", [Online]. Available: http://en.wikipedia.org/wiki/Depletion_region [Accessed 12 June 2014].

[19] S. H. Demtsu and J. R. Sites, "Quantification of losses in thin-film CdS/CdTe solar cells", in *31st IEEE Photovoltaic Specialists Conference*, 2005.

[20] J. R. Sites, J. E. Granata, and J. F. Hiltner, Losses due to polycrystallinity in thin-film solar cells. *Solar Energy Materials and Solar Cells* 55, 1–2, 43–50, 1998.

[21] B. G. Streetman and B. Sanjay, *Solid State Electronic Devices*, Upper Saddle River, NJ, Prentice Hall, 2000.

[22] First Solar, "Agua Caliente Solar Project", [Online]. Available: http://www.firstsolar.com/en/about-us/projects/agua-caliente-solar-project [Accessed 11 June 2014].

[23] NREL, "National Center for Photovoltaics", [Online]. Available: http://www.nrel.gov/ncpv/ [Accessed 10 June 2014].

[24] B. Wire, "NRG Energy and MidAmerican Solar Complete Agua Caliente, the World's Largest Fully- Operational Soalr Photovoltaic Facility", [Online]. Available: http://www.businesswire.com/news/home/20140429005803/en/NRG-Energy-MidAmerican-Solar-Complete- Agua-Caliente#.U6i4z_ldV8E [Accessed 8 June 2014].

[25] D. Swanson, R. Geisthardt, J. McGoffin, J. Williams, and J. Sites, Improved CdTe solar-cell performance by plasma cleaning the TCO layer. *IEEE Journal of Photovoltaics* 3, 2, 838–842, 2013.

[26] D. Swanson, Development of Plasma Cleaning and Plasma Enhanced Close Space Sublimation Hardware for Improving CdS/CdTe Solar Cells. Thesis, Fort Collins, CO, Colorado State University, 2012.

[27] M. A. Tashkandi, Pinholes and Morphology of CdS Films: The Effect on the Open Circuit Voltage of CdTe Solar Cells. Dissertation, Fort Collins, CO, Colorado State University, 2012.

[28] P. J. Sebastian, S. A. Gamboa, M. E. Calixto, H. Nguyen-Cong, P. Chartier, and R. Perez, "Poly-3-methylthiophene/CuInSe 2 solar cell formed by electrodeposition and processing. *Semiconductor Science and Technology* 13, 1459–1462, 1998.

22 IoT-Based Smart Hand Sanitizer Dispenser (COVID-19)

Soham Deshpande, Aakash Aggarwal,
Abraham Sudharson Ponraj, and J Christy Jackson
Vellore Institute of Technology

CONTENTS

22.1 INTRODUCTION

In late 2019, a new kind of virus, SARS-CoV-2, was found in China. SARS-CoV-2 has been recognized by the World Health Organization (WHO) as a new form of coronavirus. The epidemic has spread all over the world. COVID-19 is caused by SARS-CoV-2, which induces inflammation of the respiratory tract. It affects the upper respiratory tract (nose and throat) or lower respiratory tract of a human (windpipe and lungs). The virus is primarily spreading through respiratory droplets and contact. Hand-sanitizing and wearing a mask are the most common ways to avoid the transmission of the infection and prevent people from becoming infected. Because of this, the demand for hand sanitizers has increased. Alcohol-based hand sanitizers are applied by spraying the solvent as one presses a pump with one hand. This allows more people to come into touch with the pump's handle, which raises the likelihood of viral transmission. Also, there are dispensers where people have to work with their feet, but pressing them is irritating, and often people walk by without disinfecting their hands. In addition, the use of sanitizers varies with humans, making it impossible to estimate the level of use and handle refills and substitutes.

Because of this reason, the use of hand sanitizers is limited, which does not prevent the transmission of the virus. In the outside of some places such as restaurants, coffee shops, and grocery shops, there is a person appointed to dispense the sanitizer. He/she checks the temperature of the customers using the temperature gun and also dispenses the sanitizer manually. This is very risky as the appointed person comes in contact with many people throughout the day. These types of equipment (pressure bottles and temperature gun) are handled physically by more than one person throughout the day, leading to physical contact between them. Hence, there is an urgent requirement of a device that can do this work automatically and effectively. This device will eliminate the problem of human interference and will make hand sanitization safe and comfortable.

22.2   LITERATURE SURVEY

On December 12, 2019, the first patient suffering from coronavirus was reported. At least 1975 cases have been reported until January 25, 2020 [1]. Since then, this recent dangerous infection called coronavirus has spread at a high rate. In Ref. [1], the authors first tried to classify potential etiological agents associated with extreme respiratory diseases in the city of Wuhan. It has been reported that the infectious person has extreme respiratory syndrome, including fever, dizziness, and cough. As this is a discovery report, the number of persons is unrelated to the conclusions reached in this article. There was a tremendous increase in the need for quality health workers all over the world [2]. Healthcare workers (HCWs) are the most important players for the COVID-19 outbreak. It has been observed that SARS-CoV-2 has been transmitted from person to person through respiratory droplets and close contact, as previously documented in SARS-CoV-2 and Middle East coronavirus respiratory syndrome (MERS-CoV). The WHO has strongly advised that HCWs ask people to cover their nose and mouth with cloth, towel, or elbow when coughing or sneezing in public and include face masks and hand hygiene devices to patients suspected of having coronavirus [2]. In Ref. [3], DR. Yousaf Adam Ali says the importance of washing hands regularly. He also states that the most popular way to clean hands is to apply alcohol-based hand sanitizers regularly that claim to kill 99.99% of microorganisms. There are many sanitizers that are sold in the different forms such as liquid, gel, and foam. They are basically applied on a person's palm and then rubbed all over the hand correctly until it disappears.

The analysis in Ref. [4] indicates the value of the use of easy-to-use hand hygiene devices and how they could lead to improved staff efficiency. Proper selection of infection control devices is an important environmental intervention to improve hand hygiene. Contactless dispensers are more favored by the customer than some.

In Ref. [5], the authors designed a fully touch-free automated hand cleaning system using a Radio Frequency Identification (RFID) and infrared (IR) sensor that uses water as the medium. This system starts with spraying soap on hand and finally drying it. In Ref. [6], a communication system between different soap dispensers is created placed at different sites, which makes sure that the individual does not enter the premises without washing his/her hand. A motion sensor is used to detect the passage of the individual, and a communication network is configured among the dispensers.

Furthermore, in Ref. [7], the authors proposed an automatic touch-free hand sanitizer dispenser for COVID-19 which detects hand position using an ultrasonic sensor, and with the help of a microcontroller, Arduino UNO activates the relay that pumps the sanitizer for a fixed time and gets ready for the next action easily within 4 seconds of duration. A similar system was proposed by Akshay Sharma A S [8] in 2020. The author also mentions the places where the presence of this kind of system would be ideal. In Ref. [9], the authors have developed an automatic hand sanitizer system that is compatible with a range of containers. When the customer puts his hand close to the system sensor, the hand sanitizer is pumped by the device.

The automated hand sanitizer system suggested in Ref. [10] is essentially intended to lead to contactless hand disinfection in public spaces and prevent virus infections and is economical and eco-friendly.

In Ref. [11], the authors emphasize the importance of healthcare applications and how the Internet of things (IoT) can be used to predict diseases. Advanced instruments may be worn or inserted within the human body to monitor their health on a daily basis. Knowledge obtained in this way can be analyzed, extracted, and aggregated in order to forecast early diseases. The challenges in integrating IoT health monitoring technologies in the real world are also highlighted in the article. In Refs. [12–14], an automatic water dispenser with water level monitoring is proposed using sensors in the IoT environment. For an automatic water dispenser, the authors used NodeMCU and an ultrasonic sensor in the IoT environment. Here, the manual taps are replaced with smart taps that open and close on their own automatically. Due to this, water is saved. This dispenser not only saves water, but also sends a notification to the authorized person when the level of water becomes low through an app based on Android. Once the authorized person receives the notification for a low water level, the Android application will turn on the water supply for water cans or water tanks. In Refs. [14,15], the authors explained in detail how the body temperature is related to coronavirus symptoms. A patient suffering from coronavirus suffers from fever in most of the cases; thus, the temperature of the body will be higher than usual. In most of the circumstances, the temperature measured from the wrist is more accurate than the temperature measured from the forehead [15]. IR thermometers (IRTs) are fast, convenient, and easy to use. The findings have shown that these IRTs provide sufficient accuracy. There are many algorithms discovered until now to detect mask. One of the ways to detect face mask is semantic segmentation [15]. Training is done fully through Convolutional Neural Networks to semantically segment the faces present in the image. Binomial cross-entropy is used as a loss function, and the gradient descent is used for preparation. In addition, the output image from the FCN is analyzed to eliminate unnecessary noise to prevent false predictions, if any, and to create a quadrilateral border around the faces. Unfortunately, its accuracy to detect the face mask correctly is low as compared to other deep learning algorithms. Deep learning algorithms are the best and most accurate algorithms for face mask detection and recognition to date. In Ref. [18], a practicable solution has been suggested, which consists of, first, the identification of facial regions. The occluded face recognition issue was solved using the multi-task cascaded convolutional neural network (MTCNN). Then, the facial features are removed using the Google FaceNet embedding model and eventually classified using the support vector machine (SVM).

22.3 PROPOSED SYSTEM

The proposed method is a more practical and elegant approach to the conventional way of sanitizing hands in public areas. This product dispenses the same amount of sanitizer every time. It uses an ultrasonic sensor to detect the hand and then dispenses the sanitizer. Furthermore, there are two additional features: face mask recognition and temperature detection. Hence, whenever a person sanitizes his/her hand, the product will automatically check whether or not the person is wearing a mask and also measure his/her body temperature and simultaneously will upload the data on the cloud. As the temperature of the customer is being measured, in the future, the data can be useful for monitoring the health of the person and may also be useful for future studies. There is also an interrupt button. If the system runs into an error, the user can reset it by pressing the interrupt button. The whole system is contactless, and hence, it is very user-friendly (Figure 22.1).

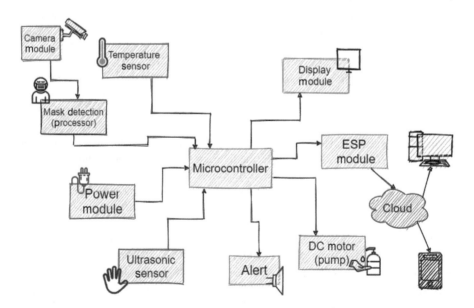

FIGURE 22.1 Schematic diagram of the IoT-based smart contactless hand sanitizer dispenser.

22.4   METHODOLOGY

The proposed system follows a procedure to generate the results. Taking an example of deploying this dispenser outside a general store, the first task it will perform is to sanitize the visitor's hand without any physical contact before entering into premises as the automatic door won't open until the dispenser commands (Section 22.4.1). Simultaneously it will perform two tasks: checking the body temperature of the visitors using an IR temperature sensor through their wrist and detecting the face mask using a deep learning-based face mask detection algorithm (Sections 22.4.2 and 22.4.3). After performing these three tasks, the visitor would be allowed to enter depending upon the temperature value and the presence of mask. Lastly, it will gather all the results and upload it on the cloud, thus maintaining a database for further use (Section 22.4.4). The defaulter may be asked to wear a mask or see a doctor as his/her body temperature is not in the expected range.

22.4.1   HAND DETECTION

Firstly, the model checks the presence of hand using an ultrasonic sensor (Figure 22.2). The ultrasonic sensor is attached to the Arduino UNO that is embedded in the chassis (the hardware model) (Figures 22.6 and 22.8). If the hand is present there in the particular range from the sensor, the motor connected to the relay and a 9-V battery will spray the sanitizer on users' hand for a particular period of time. To avoid wastage of sanitizer, the ultrasonic sensor will keep on checking the hand position until it is out of the range, and if the hand is not there, it will loop back to display "dispenser is ready" for the next. Meanwhile, it will keep a count on how many users have used the dispenser so far.

22.4.2   TEMPERATURE DETECTION

In Figure 22.3 the model measures the body temperature of the user by measuring the temperature of their wrist by using an IRT sensor (MLX90614) attached to the analogy pin of an Arduino UNO embedded on the chassis (Figures 22.6 and 22.7). This automatically happens when the user brings their hand near the dispenser for sanitizer. If the measured temperature is greater than or equal to 37°C, the alert pin (buzzer, LED, etc.) is activated, and if the model is connected to some automatic door system of the store, the user is not allowed to enter the premises without taking proper measures.

22.4.3   FACE MASK DETECTION

Thirdly, the model detects whether the user is wearing any face mask or not (Figure 22.4). This is done by using OpenCV and a CNN. To detect the face mask, a neural network is created. OpenCV is used for gray-scaling and resizing the images in the training and testing dataset. After that, the face is detected OpenCV is used to crop the face out of the grayscale image. After getting the faces using the open-source Keras, the neural network is trained using the training dataset. Once the model

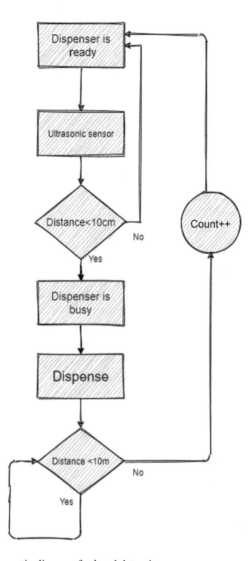

FIGURE 22.2 Schematic diagram for hand detection.

is trained, it is tested using the test dataset. A total of 20 models were built, and the one with the best accuracy and the least validation loss was used. For the application, again OpenCV was used to access the webcam, and classification (whether or not the person is wearing a mask) was done.

22.4.4 Data Aggregation (Cloud)

To analyze the types of users visiting, a record is maintained consisting of how many users have visited on a particular day, what was their body temperature, and whether they were wearing a face mask or not. The data aggregation of the system is neatly

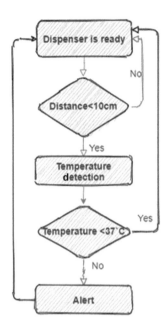

FIGURE 22.3   Schematic diagram for temperature detection.

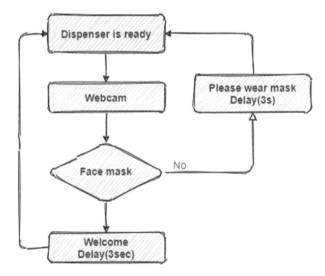

FIGURE 22.4   Schematic diagram for face mask detection.

shown in Figure 22.5. The data is automatically sent to the cloud of an open-source third-party application, ThingSpeak, via a local Wi-Fi network through an ESP module (ESP8266) attached to the Arduino UNO inside the chassis. This happens for all the users engaged with the dispenser. If a user is unfortunately infected, this database will help to backtrack and test those other users who visited before or after the

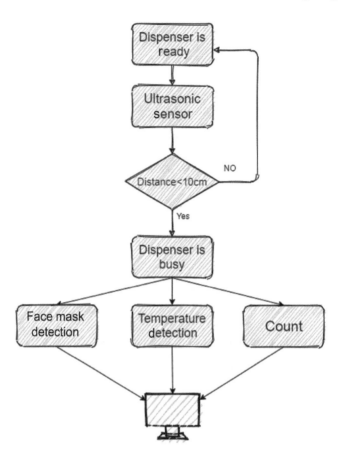

FIGURE 22.5   Schematic diagram for data aggregation.

infected person because they might have gotten in contact with him/her. Thus, this will help to keep the spreading of disease under control.

22.5   IMPLEMENTATION

22.5.1   HARDWARE IMPLEMENTATION

Arduino Uno: The proposed device uses Arduino Uno, a microcontroller board based on ATmega328P. It has 14 optical input/output pins and 6 analog inputs, and a 16-MHz ceramic resonator (CSTCE16M0V53-R0), a USB link, a power port, an ICSP header, and a reset key. The 0–12 digital pins and A0 (analog) pin were used to read the value of the temperature sensor. The Arduino Uno serves as a central core for all of the modules. In this unit, it serves as an interface that links all other components such as the ultrasonic sensor, LCD, temperature sensor, and the ESP8266 module. RTOS is not needed for this device; hence, its On-Chip RAM and ROM are sufficient to carry out the tasks. Arduino Uno, with its performance and low cost, is a better fit for the proposed device.

Ultrasonic Sensor: The proposed device uses ultrasonic sensors to track the palm. Since they are not influenced by smoke or other invisible materials, they are favored over IR sensors. After configuring the ultrasonic sensor, the pulseIn() feature is used to read the value (travel time) and assign the value to the duration vector. Using this function, the ultrasonic sensor will measure the distance between the customer's hand and the nozzle. The threshold was set for 25 cm. Hence, whenever a person brings his/her hand in the range of the ultrasonic sensor, the dispenser dispenses the sanitizer.

ESP8266: ESP8266 is a Wi-Fi module that has a system on a chip incorporated with the TCP/IP protocol stack that can allow any microcontroller access to a Wi-Fi network. It is also able to host and unload all Wi-Fi networking functionality from another application processor. It is preprogrammed and comes with the firmware package for the AT command. So, anybody can just attach it to the Arduino system and get the Wi-Fi ability. Hence, using this, all the real-time data of the proposed system are uploaded to the cloud.

MLX90614: MLX90614 is a digital temperature sensor using a contactless IR sensor. It is used to monitor and measure body temperatures ranging from 70°C to 382°C. It uses IR rays so that it can measure body temperature without any physical interaction, and it interacts with the microcontroller using the I2C protocol. It is a contactless sensor with very high accuracy. Thanks to its high accuracy and precision, it is often used in a wide variety of industrial, health, and household applications such as room temperature control and body temperature assessment.

Figures 22.6–22.8 show a custom-designed 3D-printed model of the dispenser. It has dedicated places for different components of the dispenser: a camera, nozzle, ultrasonic sensor, and temperature sensor.

FIGURE 22.6   3D-printed model of the dispenser with a camera, ultrasonic sensor, temperature sensor, and nozzle.

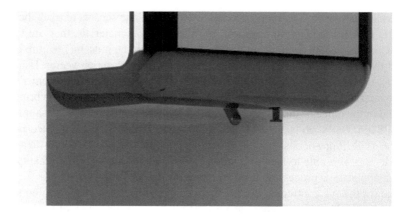

FIGURE 22.7   Nozzle of the dispenser and temperature sensor.

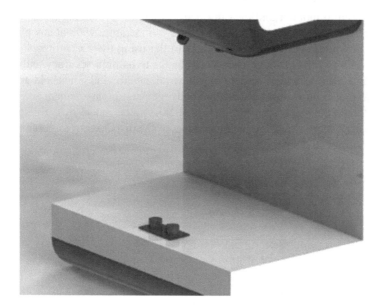

FIGURE 22.8   Ultrasonic sensor.

22.5.2   SOFTWARE IMPLEMENTATION

Face Detection: For face detection, OpenCV is used. It provides the basic infrastructure for computer vision applications. To convert the images into grayscale images and also to resize the images into the size of 100×100, OpenCV is used. Hence, first all the images are converted into 100×100 pixels. Simultaneously they are converted into grayscale images as color is not important to detect the face mask. Figure 22.9 shows this process. After resizing the images, the images are normalized by dividing their size by 255 to convert them into 0 and 1.

100 X 100

FIGURE 22.9   Colored image converted to a grayscale image.

These images then are used to train the neural network. As the neural network needs a four-dimensional array, dimensional information of each image is converted to a four-dimensional array by adding 1 as these are grayscale images (Figures 22.10 and 22.11).

22.5.2.1   CNN Architecture

A CNN is used to decide whether a person is with a mask or without a mask. Two convolutional layers are used: the first one having 200 (3×3) kernels and the second one having 100 (3×3) kernels. Then, these layers are flattened, and hence, a new layer is created. A dropout layer is used to reduce the overfitting. Finally, it is connected to a dense layer of 50 neurons, and at the last, the output layer has two neurons (with a mask and without a mask). After the first and second layers, a rectified linear unit (ReLU) layer and a pooling layer are used. The ReLU layer is a piecewise linear function that acts as the activation function. It gives the output directly if the input is having a value greater than 0, i.e., positive, and the pooling layer is used to reduce the number of parameters and computation in the network.

22.5.2.2   Face Mask Detection Algorithm Using CNN

Face Mask Detection Algorithm

1	**Data Preprocessing**
2	Import libraries Cv2, OS, Numpy, and Keras
3	**For** every image in the dataset
4	Convert images to gray scale using cv2.cvtColor()
5	Resize the images to 100×100 using cv2.resize()
6	Append the resized images to the dataset
7	Create a target list with 1 (with mask) and 0 (without mask)
8	Normalize the data variable by dividing every value by 255
9	Reshape data into 4D by adding 1 to the data array by using np.reshape()
10	**End For**
11	**Building CNN**
12	Adding ReLU layer, i.e., the activation layer, and the pooling layer
13	Adding the flatten layer and then the dropout layer to decrease the overfitting

(Continued)

	Face Mask Detection Algorithm
14	Adding the dense layer and finally the last layer with two neurons (with mask or without mask)
15	Split the data into 20% testing and 80% training
16	**Detecting faces with and without mask**
17	Import cascade classifier for face detection
18	Using cv2.VideoCapture(), capture the video from the camera
19	**While(TRUE)**
20	Reading frame by frame from the camera using source_read()
21	Converting the images to gray scale and then scaling them down to get the face
22	Resizing the image 100×100 and then normalizing the image
23	After resizing, reshaping the image to get a 4D array
24	Normalizing by dividing resized images by 255
25	Prediction using the trained model
	If 0 is having higher probability, then with mask, and if 1 is with higher probability, then without mask
26	Plot the border rectangles for the face using cv2.rectangle()
27	Destroying the windows and releasing the source
28	**End While**

The whole dataset is split into testing dataset and training dataset, 30% and 70%, respectively. After the splitting, the models are trained. Twenty-five models (epochs) were trained. After compiling the models, checkpoints were created to save the models. The models having validation loss greater than a specific value were chosen, and hence, the best models were saved.

As mentioned earlier, 25 models (epochs) were trained. In Graph 1, training loss is the loss or the inaccuracy while training the model. Hence, as more and more number of models were trained, this value kept decreasing. On the other hand, validation loss is the inaccuracy while testing. The validation loss is more important than the training loss. Figure 22.12 shows that for the 9th, 17th, and 19th models (epochs), the validation loss is the least compared to other values.

Also, in Graph 2 the training accuracy and the testing accuracy are plotted. When the models were compared, the 17th model has the maximum validation accuracy or testing accuracy (94.93%; Figure 22.13). The execution time for the model was 1 minute 214 milliseconds per step with a loss of 0.1887. Figure 22.14 shows the output.

So model 17 was chosen for implementation purposes.

22.5.2.3 Data Aggregation and Visualization

The tool used to aggregate and visualize data is the open-source ThingSpeak platform. ThingSpeak™ is an IoT analytics software service that helps users to compile, visualize, and interpret live data sources in the cloud. ThingSpeak provides instant visualization of data submitted to ThingSpeak by the user's computers. The program has the ability to execute MATLAB® code and can perform online data analysis

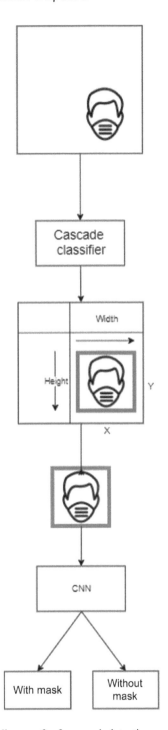

FIGURE 22.10 Schematic diagram for face mask detection.

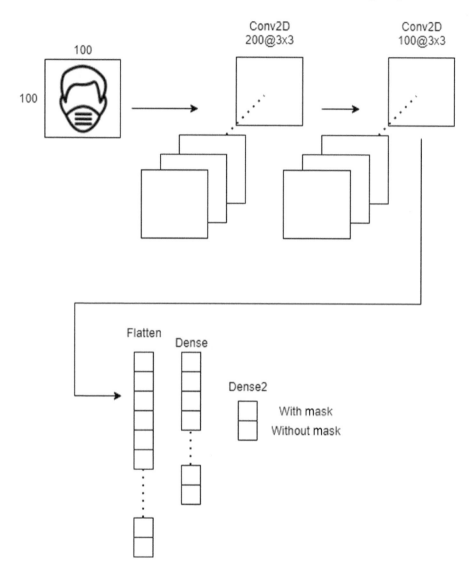

FIGURE 22.11 Schematic of the convolutional neural network (CNN).

and manipulation as it receives. ThingSpeak is also used for prototyping and proof of IoT concept structures that require analytics (Figure 22.15).

Figures 22.16–22.18 show the outputs as seen on the ThingSpeak platform (1 for with mask and 0 for without mask).

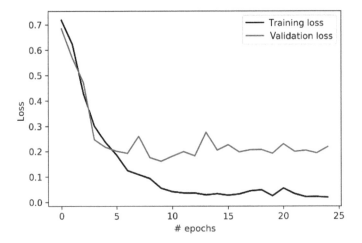

FIGURE 22.12   Epochs vs. loss.

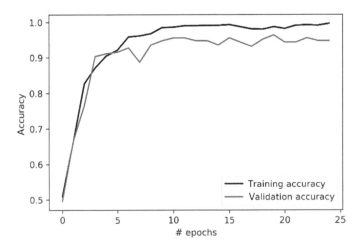

FIGURE 22.13   Epochs vs. accuracy.

Table 22.1 shows the status of eight users using the dispenser in serial order.

The red-row users are either not wearing the mask or not having a normal body temperature and thus they are restricted to enter the premises, whereas the green row indicates that the users are allowed.

No mask
https://api.thingspeak.com/update?api_key=3E8CV2EHTC146PKQ&field1=0
<http.client.HTTPResponse object at 0x7f16d45b0048>

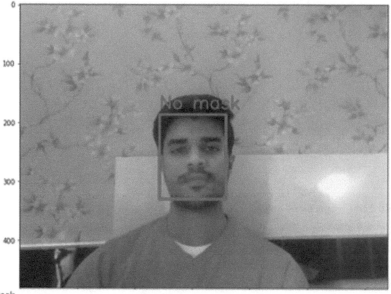

Mask
https://api.thingspeak.com/update?api_key=3E8CV2EHTC146PKQ&field1=1
<http.client.HTTPResponse object at 0x7f16d4cdce80>

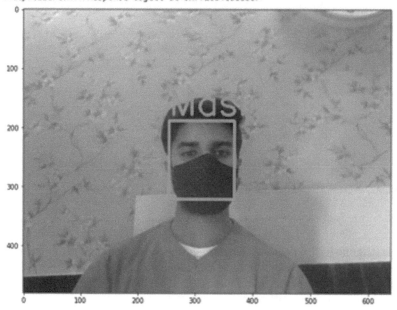

FIGURE 22.14   Face mask detection.

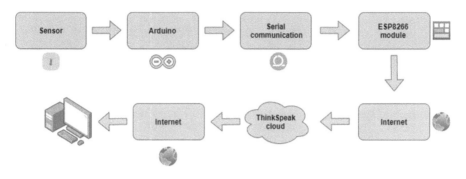

FIGURE 22.15 Block diagram for sending data to ThingSpeak.

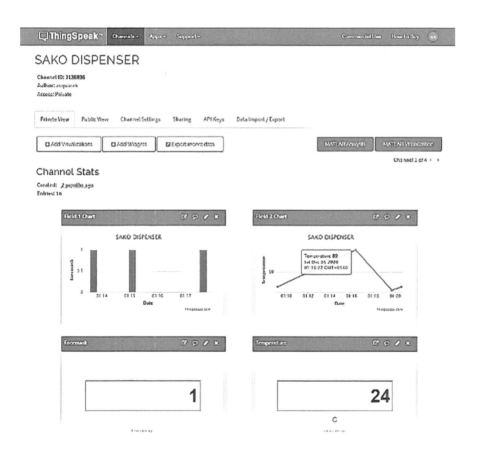

FIGURE 22.16 ThingSpeak platform screenshots with the channels.

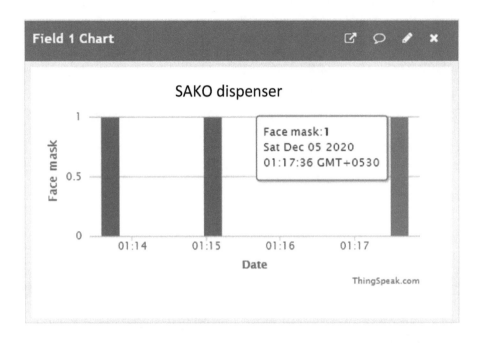

FIGURE 22.17   Status of the mask.

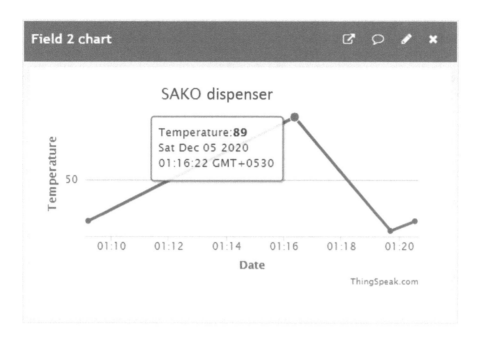

FIGURE 22.18 Body temperature.

TABLE 22.1

User Status Stored in the Database

Name	Ph. No.	Date and Time	Entry ID	Face Mask	Temp (°F)
Soham Deshpande	1234567890	2020-12-04 19:34:08 UTC	1	1	98
Aakash Aggarwal	1234567890	2020-12-04 19:35:54 UTC	2	0	100
Shubham Patil	1234567890	2020-12-04 19:39:12 UTC	3	1	97
Swara Deshpande	1234567890	2020-12-04 19:40:28 UTC	4	1	98
Shreshth Vats	1234567890	2020-12-04 19:40:58 UTC	5	1	104
Aanchal Saraswat	1234567890	2020-12-04 19:41:36 UTC	6	1	99
Millena Jena	1234567890	2020-12-04 19:42:07 UTC	7	1	98
Shreya Mahna	1234567890	2020-12-04 19:43:16 UTC	8	1	89
Naman Sharma	1234567890	2020-12-04 19:45:12 UTC	9	0	99

22.6   CONCLUSION AND FUTURE WORKS

A brand-new system has been developed having four applications, the primary one being the automatic sanitizer dispenser and the others being temperature detection, face mask detection system, and data aggregation system. This will lead to a consistent and healthy sanitizing experience for both the customer and the store owner. The smart hand sanitizer dispenser sanitizes the customer's hands without any physical contact. It eliminates the need for a person to sanitize customers' hands. Hence, it also reduces the risk of disease transmission. The IoT-based system consists of an Arduino Uno, and several other secondary components are linked to it, such as a camera, temperature sensor, and ESP8266 Wi-Fi module. The ultrasonic sensor and the temperature sensor are mounted on the body of the dispenser. As the customers' hands are detected by the ultrasonic sensor, the dispenser dispenses the same amount of sanitizer every time. Simultaneously it checks the hand temperature and looks for the mask on the customers' face. All this information is uploaded on the cloud. If the face mask is not detected, the door to the shop is not opened; in other words, the customer is not allowed to enter the shop. The cloud data can be accessed by a mobile or a computer, and it can be used for further studies. If there is any problem with the dispenser, the interrupt button can be used to reset the machine.

Owing to lack of time and experience, several diverse modifications, studies, and experiments have been left for the future. Future studies concern a closer study of existing processes, new ideas to explore alternative approaches, or merely curiosity.

The following are a few areas for future research:
- A new feature can be added to check social distancing among users.
- The device can also do face recognition of users and can only permit those users who are allowed to enter into the premises.
- The barcode scanner can be attached with the chassis to scan the user's Aarogya Setu pass.

- The information collected about the user or customer can be further used to analyze and can also be sent immediately to the nearest medical store for further aid.
- Also the name and number of the user can be added into the database.

ACKNOWLEDGMENT

We would like to thank and acknowledge the contribution of Mr. Shubham Patil (patilshubham1204@gmail.com) in the 3D design of the proposed sanitizer dispenser system shown in Figures 22.6–22.8.

REFERENCES

[1] F. Wu, S. Zhao, B. Yu, Y.-M. Chen, W. Wang, Z.-G. Song, ... Y.-Z. Zhang, (2020). A new coronavirus associated with human respiratory disease in China. *Nature* 579, 265–285.

[2] N. Lotfinejada, A. Peters, & D. Pittet, (2020). Hand hygiene and the novel hand hygiene and the novel healthcare workers. *Journal of Hospital Infection* 105, 776–777.

[3] D. Y. Ali, (2015). To study the effect of hand sanitizers used in kingdom of Saudi Arabia against the common bacterial pathogens. *International Research Journal of Natural and Applied Sciences* 2, 16–28.

[4] E.L. Larson, Albrecht, S., & O'keefe, M.L. (2005). Hand hygiene behavior in a pediatric emergency department and a pediatric intensive care unit: comparison of use of 2 dispenser systems. American Journal of Critical Care: An Official Publication, American Association of Critical-Care Nurses, 14, 4, 304–311.

[5] J. E. Pasquarelli, A. G. Bianchi, D. J. Markham, and J. T. Spring, (2013). All-in-one hand washing system, Worcester Polytechnic Institute.

[6] W. Ferniany, (2018). Systems and methods for encouraging hand washing compliance (U.S. Patent No. US 9, 940, 819 B2) UAB Research Foundation.

[7] L. Rakshith, & K. B. ShivaKumar, (2020). A novel automatic sanitizer dispenser. *International Journal of Engineering Research & Technology (IJERT)*.

[8] A. S. Sharma, (2020). Review on automatic sanitizer dispensing machine. *International Journal of Engineering Research & Technology (IJERT)*.

[9] J. Lee, J.-Y. Lee, S.-M. Cho, K.-C. Yoonu, Y. J. Kim, & K. G. Kim, (2020). Design of automatic hand sanitizer system compatible with various containers. *Healthcare Informatics Research* 26, 3, 243–247.

[10] M. Satya, S. Madhan, & K. Jayanthi, (2018). Internet of things (IoT) based health monitoring system and challenges. *International Journal of Engineering & Technology* 7, 175–178.

[11] A. Narayan, A. Krishnan, S. Ragul & A. S. Ponraj, (2020). "IOT Based Comprehensive Retail Malpractice Detection and Payment System*." 2020 IEEE International Conference on Electronics, Computing and Communication Technologies (CONECCT), 1–8.

[12] T. Anuradha & S. Jadhav, & S. Mahamani, (2019). Smart Water Dispenser and Monitoring Water Level in IoT and Android Environment. *International Journal of Computer Sciences and Engineering* 7, 810–814. doi:10.26438/ijcse/v7i5.810814.

[13] M. Parashar, R. Patil, S. Singh, V. VedMohan, & K. S. Rekha, (2018). Water level monitoring system in water dispensers using IoT. *International Research Journal of Engineering and Technology (IRJET)* 1217–1220.

[14] "Enforcement Policy for Telethermographic Systems During the Coronavirus Disease 2019 (COVID-19) Public Health Emergency", *Center for Devices and Radiological Health, Office of Product Evaluation and Quality*, Docket Number: FDA-2020-D-1138, 2020.

[15] H.-Y. Chen, A. Chen, & C. Chen, (2020). Investigation of the impact of infrared sensors on core body temperature monitoring by comparing measurement sites. *Sensors* 20(10).

[16] G. Chen, J. Xie, G. Dai, P. Zheng, X. Hu, H. Lu, L. Xu, X. Chen, & X. Chen, (2020). Validity of wrist and forehead temperature in temperature screening in the general population during the outbreak of 2019 novel coronavirus: a prospective real-world study. *medRxiv.*

[17] T. Meenpal, A. Balakrishnan, & A. Verma. (2019). Facial mask detection using semantic segmentation. *International Conference on Computing, Communications and Security (ICCCS)*, IEEE, pp. 1–5.

[18] M. Ejaz & M. Islam, (2019). Masked face recognition using convolutional neural network, 1–6. doi:10.1109/STI47673.2019.9068044.

[19] Y. Wan, (2013). Tracking and hands motion detection approach for monitoring hand-hygiene compliance for food handling and processing industry.

Index

Note: **Bold** page numbers refer to tables and *italic* page numbers refer to figures.

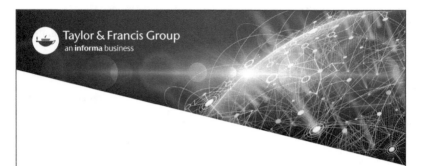